THE BIG HOP

ALSO BY DAVID ROONEY

About Time:
A History of Civilization in Twelve Clocks

THE
BIG
HOP

THE FIRST NONSTOP FLIGHT
ACROSS THE ATLANTIC OCEAN
AND INTO THE FUTURE

DAVID ROONEY

W. W. NORTON & COMPANY
Independent Publishers Since 1923

For information about permission to reproduce selections from this book, write to
Permissions, W. W. Norton & Company, Inc., 500 Fifth Avenue, New York, NY 10110

For information about special discounts for bulk purchases, please contact
W. W. Norton Special Sales at specialsales@wwnorton.com or 800-233-4830

Manufacturing by Lakeside Book Company
Book design by Patrice Sheridan
Production manager: Ramona Wilkes

ISBN 978-1-324-05096-4

W. W. Norton & Company, Inc., 500 Fifth Avenue, New York, NY 10110
www.wwnorton.com

W. W. Norton & Company Ltd., 15 Carlisle Street, London W1D 3BS

10 9 8 7 6 5 4 3 2 1

This book is dedicated to

JOHN "JACK" ALCOCK

CONRAD BIDDLECOMBE

ARTHUR WHITTEN "TED" BROWN

HARRY HAWKER

KENNETH MACKENZIE "MAC" GRIEVE

CHARLES MORGAN

FREDERICK "FRED" RAYNHAM

and all the other aviation pioneers—
men and women, of all ages, from all walks of life—
who risked everything:
for freedom, for progress, and for us.

CONTENTS

PART THREE

THE BIG HOP

PROLOGUE

HARRY HAWKER AND MAC GRIEVE HAD BEEN FLYING EAST OVER the Atlantic Ocean for ten hours when the end came.

The flight had started in rocky, snow-covered Newfoundland well enough. The men had planned the long flight carefully. After a slightly shaky take-off, when they nearly crashed into a ditch, they made good progress.

But soon, things started to go wrong.

They hit fiercer storms than the meteorologists had predicted. Nobody really knew what the weather was like a mile or two above the mid-Atlantic because nobody had been there before. As Hawker called on the aeroplane's engine for more power so he could skirt around the worst of the storm, he found that it had developed a fault. It wasn't a fuel shortage, nor a mechanical malfunction; there was a problem with the cooling system. The water in the radiator wasn't getting around the engine, so it was overheating. The men tried to cure the problem in a series of risky manoeuvres. But it was no good: the coolant was boiling off fast, and when it ran out, the engine would soon seize.

They came down low over the roiling Atlantic waters to look for somewhere to ditch, and for a ship that might pick them up. Two hours of searching, as the radiator's pressure-relief pipe spewed steam in front of them, proved fruitless.

Harry Hawker and Mac Grieve in Newfoundland, ready to attempt the Big Hop.

The thirty-year-old Hawker had years of experience and was considered one of the best pilots in the world, but he was highly strung. When things were going wrong, he could be moody and impatient. He struggled with the controls as the storm tossed the aeroplane around the sky. It was like cresting a rollercoaster repeatedly, and the pit of his stomach heaved each time. Before long, he was vomiting over the open side of the cockpit. Grieve, the navigator, was a navy man, nine years older than his partner and more stolid. He had been trying to work out their position with his sextant, drift indicator, and compass. But the rain squalls and dense clouds had cut him off from the sun and stars, even if he could have held the sextant still enough as the aircraft lurched. In the end, he joined Hawker in scouring the sea for salvation.

They flew back and forth in a zigzag, but the scene—what little of it they could glimpse as they stared into the storm—never changed. The sky was theirs and theirs alone. The ocean, a turbulent grey waste 400 feet beneath their cockpit, was hiding the steamships that ploughed its frigid waters. The wireless set that might have alerted a ship to their distress hadn't worked all through the flight. The men had flares on board: bright red lights that could be fired high into the sky from a pistol. But they would be useless in fog like this. The cloud was too thick, and the flares would sputter out in moments anyway.

The two aviators had flown a thousand miles from Newfoundland in their two-seater aircraft with its open-topped cockpit. They'd have to fly another thousand to reach where they planned to go: Ireland.

The wind and spray whipped their faces raw. The harsh growl of the engine assaulted their ears and the heady fumes of gasoline exhaust flowed around their cramped enclosure. They knew that millions of people around the world were willing them along on their flight. But the men had never felt more alone. They were now resigned to the fact that they would not make it across the ocean. There they were, at eight o'clock that morning: a speck of humanity waiting to die in an angry Atlantic storm. Finally, the last drops of precious coolant in their engine boiled away.

IT WAS MAY 1919. HAWKER AND GRIEVE WERE AMONG SEVEN AVIAtors that summer who would take off from Newfoundland, intending to land on the shores of the British Isles without stopping along the way. The contest for the first nonstop transatlantic flight was sponsored by Britain's *Daily Mail* newspaper. It was known to some, especially in the United States, as "the Big Hop." Only two of the seven airmen would make it across successfully. The others would fail in their endeavours.

Yet they all achieved something which, at the time, was akin to greatness. Even before they went out to Newfoundland in the spring of 1919, each of the seven airmen possessed a life story that could fill a book or a movie screen. They ranged in age from twenty-five to thirty-nine. Three of them were to act as pilots, and four would be navigators; all seven had trained to fly at one time or another. Each had experienced more in his short life than most people could comprehend.

As the events of 1919 unfolded, the contestants were feted on both sides of the Atlantic by a public hungry for stories of success. The terms

Mac Grieve, Fred Raynham, Harry Hawker, and Charles Morgan, four of the contestants getting ready in Newfoundland to fly the Atlantic in 1919.

of peace after the Great War had not yet been agreed. The influenza pandemic, its waves still washing around the globe, would take many more lives than the war. The transatlantic flight contest offered a welcome distraction. But the public adulation that summer faded fast. The story of the Big Hop was quickly forgotten once the contest was over. Those who took part were modest and did not seek glory. The public, for their part, realized they were weary of seeing the dangerous consequences of aviation. Families had lost enough young aviators in the war to know that flight could be a world of pain as well as promise. Perhaps the contest didn't feel so glorious in the end.

Some of that pain faded a little as the 1920s wore on. Civilian air travel gradually established itself, at least over short distances. Aircraft and their engines improved, and flying became more of a fixture in the world as its worst dangers receded from view. It was this later world, not the one just emerging from the war, that was truly ready to celebrate transatlantic flying. It was an eighth airman, the American Charles

Lindbergh, who first flew the Big Hop alone. He made his solo transatlantic flight in 1927, and the eight years that had passed since 1919 made a lot of difference. His achievement made him a lasting celebrity. Bestselling books and a blockbuster movie kept his accomplishment in front of the public's eyes for decades. In 1932, Amelia Earhart made the second nonstop solo crossing, and she caused a sensation as well. These later successes only pushed the story of 1919 still farther into the shadows. Many today still think Lindbergh was the first to fly the Atlantic.

The seven airmen of 1919 each learned to fly on flimsy aeroplanes made of timber struts and varnished fabric; aeroplanes which could carry two people at the most. By the time the last of the seven men died, in November 1970, Boeing 747 airliners had been in service across the Atlantic for almost a year. Each of these jumbo jets at take-off weighed 300 tons. They carried upwards of 400 passengers at 600 miles per hour in luxurious comfort. It wasn't long before such flights became commonplace, then mundane.

Today, a transatlantic flight is an unremarkable part of everyday life. It is almost a chore. But somebody had to go first.

PART ONE

They are heroes, modern ones, who can compare with those of Greek, Norse, and Roman fame, and their deeds are making history. Their lives are so various before they have taken up their calling that one cannot account for their choice. It mostly comes suddenly, and with such force that every obstacle is overcome to reach the desired end.

—HILDA HEWLETT, AVIATOR, 1917

CHAPTER 1

FREEDOM AND EXHILARATION

ON A QUIET SATURDAY EVENING IN MARCH 1911, IN THE SMALL Australian ranchers' town of Caramut, an intense twenty-two-year-old garage mechanic found himself in a hotel dining room, telling his friends why he was leaving for England.

Harry Hawker was born in January 1889 in Moorabbin, a township set among orchards and market gardens a few miles outside Melbourne. His father, George, was of Cornish extraction. In another life George Hawker might have become a miner. But the Victorian gold rush had slowed to a crawl before he was old enough to swing a pickaxe and dig for his fortune. Instead, he set up as a blacksmith and wheelwright, building wagons for the Chinese market gardeners transporting their fruit and vegetables along the Nepean Highway to Melbourne every day. He built his own steam engine for the workshop, and his own car, using a lathe driven by foot pedals and a pillar drill powered by hand. When Harry came along, the third of four children, he soon absorbed his father's obsession with mechanism. He once told a magazine reporter that he was "crawling under machinery from the age of three."

Harry's childhood was a regimen of strict discipline and perpetual activity. His father was a prize-winning rifle shot and served as a sergeant in the local militia. He was also a Methodist missionary who preached that idleness was a sin. "Don't stand around doing nothing," he would admonish his children; "do something, even if it is wrong." Every Sunday morning and evening, Harry and the family would worship at the local Methodist church. The rest of the time they would be working. They kept cows for milk, chickens for eggs and meat, and horses to ride. They grew their own vegetables, made their own butter, and baked their own bread. Such an active home life left Harry with little appetite for academic study. School was a place he tolerated each day until he could escape its confines and return to his tools. He left school in 1901, aged twelve.

Hawker spent his early teens hopping around Melbourne from one motoring job to the next. Fired with the mechanical zeal of his father, he chased any opportunity to work with the latest cars, the fastest motorcycles, and the most powerful engines. He was soon able to extract the finest performance from any petrol motor and developed a reputation as one of the best mechanics around. "He thinks in gears, wheels, bores and strokes," his brother-in-law once observed. People also started to realize he was something of a maniac for speed.

At seventeen, Hawker took up work in the rural township of Caramut, 150 miles from Melbourne. It was a remote pastoral settlement for sheep and cattle ranchers with a population of just two hundred. After a couple of years, he picked up a position working for Ernest de Little, a polo player and former first-class cricketer who had taken up pastoral farming there. He developed a nice little life. There was the spacious workshop, with its modern lathe, its tidy workbenches, and whatever tools Hawker wanted. There was the Rolls-Royce Silver Ghost that he drove for de Little, who collected fine motor cars but didn't want to drive or fix them himself. There was the generous salary, paid by de Little for a job that was hardly full time for somebody with Hawker's talents. His lodgings at a comfortable Caramut hotel, run by kind and indulgent owners, were covered. And he could run his own engineering business on the side. But he was growing restless. He wanted more than rural Victoria could offer somebody like him: he wanted to escape.

In March 1910, he experienced an epiphany when he witnessed Australia's first aeroplane flights. They were made by the famous escapologist Harry Houdini over dry grassland at Diggers Rest, twenty miles outside

Melbourne. It's likely that the twenty-one-year-old Hawker helped over-haul Houdini's engine when it developed a fault. The young Australian certainly discerned in the episode a way out of his small-town existence. After one flight, Houdini told a reporter, "As soon as I was up all my muscles relaxed, and I sat back, feeling a sense of ease. Freedom and exhilaration, that's what it is." Hawker wanted in. He wasn't prepared to settle for a life constrained by status and geography. He wasn't content with the cards he had been dealt. He must have taken the view that if you stood still in the fast-changing modern world, you would take root, stagnate, and die inside.

Harry Hawker was a short young man, considered good-looking by his friends, who dressed in smart double-breasted jackets, high-waisted trousers tapering to the ankles, and shoes with high heels to elongate his diminutive frame. He had a tight, angular body, a mop of curly black

Harry Houdini flying over Diggers Rest, Victoria, in March 1910, with Harry Hawker watching.

Harry Hawker in 1912, aged twenty-three.

hair, and eyes that seemed to burn with restless energy. One writer said, "In his black Celtic eyes glow fires of Celtic enthusiasm," and that "his jaw has the firm line which spells 'grit.'" He favoured actions over words, with admirers describing him as "manly and sporting." Like his father, he took up the rifle and became a crack shot. He was a keen and fast boxer who could, as one friend recalled, "hit like the kick from a mule." In conversation he had a tendency to joke around or become argumentative. Long discussions bored him. His friends soon learned to spot the telltale sign of impatience: a quick shake of the head to toss the thick curls off his forehead. It was as if, said an acquaintance, "he wanted more air." When he did have something to say, Hawker spoke economically, and people tended to listen.

In the dining room of the Western Hotel in Caramut in 1911, he seemed to hold the gathered well-wishers in a spell as he told them why, after five years in the township, he had decided to leave. He was heading for England "to become further acquainted with motor science," he reportedly said, "and to study aviation." He was never short of Aussie optimism. He confidently assumed he would just walk into a job.

But when he reached the old country, the bruising reality of the situation hit him hard. Work was difficult to come by. He was an outsider, brash and confrontational, with no references and little inclination to flatter and plead. It was a humiliation. In letters home, Hawker fumed about the "Pommy bosses" who rejected his applications. But he persevered. His first job, after two months of searching, was with the Commer commercial vehicle company in Luton, twenty-five miles north of London. Its pay was low, so, after six months, he took up a slightly better-paid position at the London agent for Mercedes cars. He only stayed there two months before quitting for the nearby Austro-Daimler company, which offered another small increase in pay. He was relieved to be in work, but weekly visits on his days off to watch the flying at the nearby Brooklands aerodrome and racetrack only cemented his ambition to join the aviators competing for the freedom of the open skies.

In June 1912 Hawker heard that a young man called Tom Sopwith was looking for a mechanic to work at his Brooklands flying school. Sopwith was just a year older than Hawker. Like the Australian, he had grown up with a fascination for mechanism and an aversion to academic study. When he reached his teens, Sopwith planned to join the Royal Navy, but they turned him down. "They didn't think I was clever enough, and they were probably right," he later commented. Instead, he attended an engineering college on England's south coast, where he immersed himself in the practical study of motor vehicles. In 1905, at the age of seventeen, he left. A sizeable inheritance had left him comfortably well-off. With a childhood friend, he started selling motor cars in London, and business went well. He took up yachting and competed in speedboat contests on the side. He also raced motor cars at Brooklands and elsewhere, winning trophies and surprising some trackside spectators who thought the serious-looking driver with the baby face could barely have reached his teens.

At eighteen, Sopwith took up ballooning. He loved the challenge of it, but found it a capricious pastime. He nearly lost his life on one flight when high winds pushed the craft along so quickly that he was almost past England's west coast and over the Atlantic before he could bring it down to land. Then, just like Hawker, he had an epiphany. In 1910, aged twenty-two, he spent £5 on a short aeroplane joyride at Brooklands. He found the sense of space and freedom in the open skies euphoric. With the funds at his disposal, and with his relentless have-a-go attitude, he bought his own aeroplane, taught himself to fly it, and passed his tests. Then he spent two years travelling the world to compete in flying contests. He won more than he lost and made good money from the prizes on offer. Before long, he realized there could be a business in aviation. In February 1912 he set up his Brooklands flying school, and later that year, he branched out into aeroplane manufacture at the nearby town of Kingston.

When Harry Hawker turned up to enquire about the position for a school mechanic, he was in luck. Years later, the story went that Sopwith told Hawker he only wanted shop boys. Hawker reportedly replied, "All right, I will sweep the shops!" Whatever was said that day, Hawker got the job. He quickly demonstrated his prodigious technical skills. He also started to show his old swagger. A friend recalled that he had "soon made it clear that he was to be more than just another cog in the wheel," though his assertive attitude could sometimes create friction with his new workmates. Before long, he asked for flying lessons, and Sopwith agreed. Within a month, Hawker passed his tests.

Now that he was a qualified pilot, he wanted to test the extent of his talent. Tom Sopwith's experience proved that reputations and money could be made in flying contests. Hawker decided to follow suit. An endurance competition funded by the Michelin tyre company, which offered an impressive trophy and £500 to the pilot who could fly longest without touching the ground, looked like the perfect opportunity to prove his mettle. It might also be a financial lifesaver. For aviators from modest backgrounds, the money put up in contests was what allowed them to keep flying. The Michelin competition closed at the end of October. Hawker decided to take the prize for the Sopwith company.

Tom Sopwith in his aeroplane, about 1910.

THICK FOG COVERED THE BROOKLANDS FLYING VILLAGE ON THE morning of Thursday, October 24, 1912. At least it dampened the stench from the sewage farm, but that was about all it had going for it. Following the rainstorms and high wind earlier in the week, it was looking like another bad day to fly. In weather like this, it was more likely somebody would get themselves killed than break a record.

Along the railway embankment that marked the north-west boundary of Brooklands, steam trains chugged through the gloom. They were carrying affluent Surrey residents, and goods from Southampton and Portsmouth, into London, twenty miles up the line. People with lodgings in nearby Weybridge and Byfleet made their way to the aerodrome by foot or bicycle. Others came longer distances, travelling in the dark early hours on motorbikes. Some could afford neither lodgings nor ground transport. They might sleep in the wooden sheds that constituted aircraft hangars, if they could get away with it. As the flying village started com-

ing to life that morning, the sound of hand tools clattering and scraping
began to rise as people got to work. In the engine workshop, the shrill
keening of lathes and milling machines started to pierce the autumn air.
The smells of lubricating oil and gasoline and sawn timber and fabric
lacquer hung heavy and intoxicating over the sheds.

Six years earlier, Brooklands had been a 330-acre tract of woodland,
marshy meadows, and poultry farms. The River Wey wandered slug-
gishly across the site, passing under the railway embankment on its way
to join the Thames. Low hills clad in pine trees rose in the east, overrun
with rabbits, which the residents of Weybridge would attempt to snare.

The vast site was part of the 4,600-acre estate inheritance of Hugh
Locke King, who lived in one of the elegant country houses close to the
railway station with his wife, Ethel. The inheritance had given them the
means to pursue extravagant pastimes. The Locke Kings became keen
Edwardian motorists. But they grew frustrated by the speed limits that
stopped motor racing from taking place on Britain's public roads. In
1906, Alfred Harmsworth, the determined forty-year-old proprietor of
the *Daily Mail* newspaper and himself a motoring devotee, encouraged
the Locke Kings to build their own private track where cars could race
without limits, and they liked the idea. They decided to turn the marsh
and woodland near their home into a purpose-built two-and-three-
quarter-mile racing circuit.

For eight months, two thousand workers living in corrugated-iron-
and-timber huts shifted 400,000 tons of earth, chopped down thirty acres
of woodland, grubbed up carpets of bluebells, diverted the river, and
demolished the two farms that tried to scratch a living on the unpromising land. Then they built vertiginous banking at each end of the site,
higher than the houses behind it. They connected the two banked ends
on one side by a half-mile-long straight that ran alongside the railway
embankment. On the other side was a gentle reverse curve. They bridged
the river in two places. Finally, 200,000 tons of concrete, to a thickness
of six inches, was poured onto the surface of the 100-foot-wide track.
Cars on this vast new circuit could race ten abreast at 120 miles per hour.

When Brooklands opened in June 1907, it attracted plenty of attention.

But attendance at race meetings rarely approached the venue's thirty-thousand-strong capacity. After a while a new group of motor enthusiasts began to come forward, and there was a growing interest among the public to watch them at work. Britain's pioneer aviators needed somewhere to build and fly their aeroplanes. They needed wide, open, level ground. For taking off, they wanted three-quarters of a mile of turf as smooth as a tennis pitch. For landing, the ground did not need to be perfect, but there needed to be a lot of it, and no trees. And they would require hangars. Small wooden sheds would suffice—they just needed somewhere to keep their tools and their small, fragile aircraft and, for those aviators hard up for funds, somewhere to bed down at night. In 1909, ground at the western end of the racetrack was flattened for the new flying village. Sheds began to go up; the first aviators moved in.

It was eight years since Queen Victoria had died. Her sixty-three-year rule had seen the world transformed. Now, in Edward's reign, everything seemed to be changing again; society's established structures seemed to be fracturing. Britain's Labour Party, formed to give voice to the working class, had been founded in 1900, in Victoria's final year. Rising demands for women's suffrage had been catalysed by the formation in 1903 of Emmeline Pankhurst's Women's Social and Political Union, which embarked on a militant campaign against the establishment. In Russia, the 1905 revolution was still convulsing. Old ways were clinging on, but modernity was washing in relentless waves over societies worldwide. The future no longer belonged only to landowners like the Locke Kings. Opportunities for advancement and fortune were opening to those born without privilege. Brooklands, despite its establishment beginnings, was destined to become one of those extraordinary locations where egalitarian innovation flourishes—a place that attracts those who want change.

It could be a thrilling place. One visitor described "the immense flat plain of the flying ground bounded by the motor track which shuts out all other view and forms the horizon, so that one feels as one imagines a fly might feel if he found himself set down on a green silk table-spread with a white border." The flying displays put on by the pioneer aviators—those who were self-taught or who had been taught overseas—attracted huge crowds, much bigger than the motor races. Spectators would line the top of the racetrack banking to get a glimpse of this exotic new activity.

Soon, the amateurs and enthusiasts were joined by business-minded aviators who set up flying schools. The first was founded in 1910 by Hilda Hewlett and Gustav Blondeau. Hewlett had been born, as Hilda Herbert, into a stifling middle-class family in south-west London. Her claustro-phobic upbringing transformed into a marriage to Maurice Hewlett, a poet and historical novelist. But the partnership had eventually run its course and was falling apart. Searching for escape, Hewlett witnessed her first aeroplane flight in 1909, at Blackpool, when she was forty-five. She later remembered, "I was rooted to the spot in thick mud and wonder and did not want to move. I wanted to feel that power under my own hand and understand about why and how." When she told her family about her newfound ambition to fly, she was met only with scorn. Her husband, the poet, insisted that women were unsuited to aviation because they didn't have the nerve. It was just, he said, "a passing and silly escapade." But Hewlett prophesied a different future. "The time will come," she said, "when every woman who can drive her own car will also pilot her own aeroplane. All women can begin on an equality with men in the study of aeronautics as it is a new science, and it is delicate and exacting work for which they are quite fitted."

The school's first pupil, thirty-five-year-old Maurice Ducrocq, signed up in August 1910. Two months and twenty-one lessons later, Ducrocq passed his pilot's certificate. He was full of praise for the way Hewlett and Blondeau taught him. In fact, he was so inspired by the Brooklands experience that he set up his own aviation business there. He occupied a shed next to another newly formed flying school, run by a company known as Avro.

By 1912 the flying village had grown to a settlement of forty wooden hangars, filled with activity. There was the buzzing rasp of thirsty and capricious piston engines. The sensation of speed and power pervaded the place. There would be the shadow of an aeroplane lurching overhead and shouts of encouragement from other aviators. Then there would be shouts of frustration as a take-off or a landing would go wrong. Often,

Hilda Hewlett with her aeroplane at Brooklands in 1911.

there would be a forced drop into the sewage farm next to the sheds. The owners of Weybridge and Byfleet boarding houses soon got used to laundering filthy overalls. It was a world apart from the rest of British society, where everybody knew their place and kept to it. The flying community at Brooklands became a place where the gentry mixed with the proletariat. Women and men worked side by side. Teenagers could chat as equals with their elders. Britons rubbed shoulders with aviators from overseas. They all came together at the airfield's café-restaurant, the Blue Bird, where they discussed their problems and pored over the latest copies of flying magazines while smoking cigarettes, eating bread and jam, and drinking endless cups of tea.

On the cool autumn morning of October 24, the fog that enveloped Brooklands soon started to burn off, and there was barely the whisper of a breeze that day. Bad weather and bad luck earlier in the week had thwarted Harry Hawker's bids for the Michelin prize. He had already been up in his Sopwith biplane three times. On his first flight, a valve spring in his hard-working engine broke after he had been three and a half hours in the air. High winds put paid to the second attempt, despite his endeavours to get above the gusts by ascending to 2,000 feet. The aircraft was tossed around like an out-of-control kite, 200 feet up and down each time. He struggled to keep hold of the controls. There was no way he could have hung on for longer. The following day, his third try was ended by a heavy rainstorm. The water got into the magneto, which created the engine's ignition sparks, and short-circuited it.

Yet, as the sun rose over the airfield that Thursday morning and pushed the fog aside, all these tribulations were behind him. Hawker assumed he would be making the attempt for the prize on his own that day. It would be fourth time lucky, he told himself, as mechanics pulled the wooden shutters from the Sopwith hangars, lit the braziers to fend off the autumn chill, and set to work preparing the aeroplane. Then his friend Fred Raynham had turned up two doors down at the Avro shed.

The flying village at Brooklands, 1913, with the Blue Bird restaurant in the second shed from the left.

———

FREDERICK RAYNHAM, KNOWN TO MOST AS FRED, WAS A TALL nineteen-year-old with dark, straight hair combed into a neat parting, and a long, youthful face that could have made him a movie star. He was a cheerful young man, though people noticed an air of reserve about him. One friend later observed, "His carefree looks hid a careful thoughtful character." There was an edge of steel to his personality as well. As the son of Suffolk tenant farmers, born in 1893, he had been just two years old when he lost his father to tuberculosis. The thirty-two-year-old James Raynham had been staying at a treatment hotel in Davos, a town high in the Swiss Alps whose pure, frigid air made it a mecca for those suffering from lung ailments. But an Alpine convalescence failed to save him. His death left his thirty-year-old wife, Minnie, to bring up Fred and his older sister alone.

The tragedy of their bereavement could not have come at a worse time. British farming in the 1890s was on its knees, and Suffolk was the county suffering by far the worst in the agricultural depression. Grain prices had fallen through the floor. The American prairies had been put under mechanized cultivation, and railways and steamships slashed the cost of transporting the harvest overseas. American wheat flooded European shores. Under the policy of free trade, Britain's cereal farmers could not compete. Agricultural workers from Suffolk and elsewhere had been fleeing in droves to the industrial cities to find work, leaving rural populations hollowed out.

For many of those left behind, prices had fallen so low it just wasn't worth cultivating the land: it cost more to grow the crop than it fetched at market. Some tried to switch to dairy farming, but their desperate condition meant that investment in new machinery and stock was difficult to secure. Change was hard and slow: too hard and too slow for many. Surviving records suggest that James and Minnie Raynham managed to navigate the worst of the depression and switched to livestock farming before James's failing health had overtaken him. Those who couldn't make ends meet simply abandoned their land. A government investiga-

Fred Raynham in 1911, aged seventeen or eighteen.

tion in 1895, the year James Raynham died, heard from one expert witness that "Suffolk can be given away to anybody who will take it. . . . The future is about as black as can be."

There was grinding, dispiriting poverty all around, and there didn't seem to be any way out of it for those on society's bottom rung. Two reformers, writing twenty years after Fred Raynham's birth, described the hopelessness that still hung over the poorest workers in rural Britain, whose earnings kept them alive and little more. Their poverty trapped them in a hollow existence. "It means," wrote the reformers, "that people have no right to keep in touch with the great world outside the village by so much as taking in a weekly newspaper. It means that a wise mother, when she is tempted to buy her children a pennyworth of cheap oranges, will devote the penny to flour instead. It means that the temptation to take the shortest railway journey should be strongly resisted. It means that toys and dolls and picture books, even of the cheapest quality, should never be purchased; that birthdays should be practically indistinguishable from other days. It means that every natural longing for pleasure or variety should be ignored or set aside. It means, in short, a life without colour, space, or atmosphere, that stifles and hems in the labourer's soul as in too many cases his cottage does his body." The Raynhams had been better off than desperate farm labourers like those. But once James died it must have been clear there was no future for the family in Suffolk. In 1896, widowed and with two young children to support, Minnie Raynham put the farm's livestock and machinery up for sale, hoped for the best, and left.

The family moved around southern England a fair bit over the next few years. As the man of the household, Fred needed to earn his keep as soon as he could. By his mid-teens, he seems to have taken on a clerical job. Then something happened that set his life on a new course. It's not clear, from the historical record, how it happened. But in autumn 1909, at the age of sixteen, he secured a trainee mechanic's job at Brooklands, working for one of aviation's pioneers, John Neale. His new role allowed him to sit alongside Neale in test flights of a self-built monoplane. It was a short-lived opportunity. Early in 1911, Neale left Brooklands, in circumstances that have become equally blurred. This could have spelled the end of Raynham's flying career. But his mechanical prowess had been noticed around the close-knit flying village as he helped Neale with his experiments. In February, he got a job with the Avro company, and with it came flying lessons. Raynham was, it seems, a natural. On May 9, 1911, he qualified for his aviator's certificate. He was just seventeen years old.

Historical sources preserve only the scantest detail of Fred's childhood and teenage life. Given the desperation of his earliest years, and the peripatetic upbringing that followed the family's flight from rural Suffolk, this is hardly surprising. The lives of the straitened working classes in those days can now usually be glimpsed only in aggregate, rather than in richly detailed personal records. In the absence of an autobiographical account or preserved public archive, we can nonetheless discern something of Fred Raynham's character. Commentators would later remark on his energy, his skill, his technical versatility, and his bravery. Fortitude in the face of adversity would have been a characteristic he had to develop as he grew from childhood into adolescence. The ability to transform as the winds of fortune shifted would have been a vital attribute for all those farming families compelled into exodus by the depression of the 1890s.

There was another word that onlookers used to describe Fred Raynham, and perhaps it explains why there is no detailed memoir of his formative years: he was modest.

It seems fair to guess that Raynham, like Harry Hawker on the opposite side of the world, craved an opportunity to deviate from the path that his class and social status had mapped for him. Life as an office clerk must have been, for him, a bloodless existence. Perhaps Raynham was looking for adventure. He also, it seems, wanted to put himself to the test: to find the limits of his skill and his bravery, and to see how far past those limits he could push. Flying was a world of freedom and exhilaration, as Harry Houdini put it in the dry grasslands of Australia when he flew there in 1910, watched by Harry Hawker. For a frustrated teenager like Fred Raynham, there could have been little else like aviation in Edwardian Britain. Somebody like him had every chance of succeeding in this new aerial adventure. There was no need for social contacts or deep pockets. In aviation, a son of Suffolk farmers—or a Melbourne garage mechanic—could rise to the top if they wanted it hard enough. And Fred Raynham, as competitive as he was modest, wanted it—badly.

Raynham quickly settled into his job at Avro. He also took on work for Tom Sopwith, whose businesses kept expanding. Much of Raynham's time was spent giving flying lessons to the queue of pupils who now lined up for tuition. In 1910 Britain's flying schools trained just eighteen aviators between them, eight of them at Brooklands. The following year, they turned out a total of 109 qualified pilots, forty-four of them learning at Brooklands. In 1912, the schools tutored 211 successful flyers. Ninety of them took their lessons at Brooklands. Fred Raynham was responsible for a strong share of this success. But for all that tuition paid the bills, he wanted to keep pushing himself onward, like they all did. The only way to advance was to compete.

The aviators of Brooklands and elsewhere organized a busy programme of races and contests among themselves. They would race in circuits around aerodromes or across the country. They would compete to see who could get off the ground the quickest. They held competitions for the pilot who could drop a flour bomb closest to a painted target, or the one who could carry the most passengers. They competed at night flying, and endurance flying, and altitude flying. Tutors would compete against their bosses. Rival schools would compete against one another. Newly qualified aviators would compete against seasoned flyers.

Hilda Hewlett, who had founded the aerodrome's first school, was just one of many who loved the Brooklands life. She later described it as being spent "amongst men, young and old, who lived with one aim, fearing not death, caring not for other things, respecting not overmuch law and regulation that stood in their way." One Sunday afternoon, a blustery wind had been blowing across the aerodrome as eight aviators had taken part in a quick get-off race organized by the aerodrome's Aero Club. Hewlett took first place in the contest, beating Tom Sopwith into third. She was delighted to win. Her twenty-one-year-old son, Francis, was there to watch her. He'd been up in the air himself that morning. His sense of pride in his mother's achievement can only have been heightened by the fact that it was she, only months earlier, who taught him to fly.

Fred Raynham also took part in the get-off contest, but he didn't even place. He would have to try harder next time. As he pushed his little canvas-and-timber aircraft over the grass towards the shed, fighting the gusts of wind that threatened to tip it onto its nose or a wing, and thinking about the cup of tea and the good-natured joshing waiting for him at the Blue Bird, Raynham was surely smiling at the thought of how his life was going. The dismal world of his teenage clerical work must have seemed a distant memory by then. He had decisively stepped off the preordained path of a Suffolk farmer's boy in an agricultural depression.

When the Michelin endurance contest came along, it had his name all over it. His new friend Harry Hawker wanted to win it for Sopwith. But Raynham resolved to grab the prize for Avro—and for the sheer fun of competing.

BY 7 A.M. ON OCTOBER 24, HAWKER HAD BEATEN RAYNHAM INTO the air, but his flight lasted only twenty minutes before he was forced to come back down. The magneto was still wet and kept cutting out. Hawker had a spare waiting for him at the Sopwith shed, but after the mechanics fitted it, they discovered that it, too, was faulty. It had to be taken off, disassembled, and repaired, and even though the clock was ticking, the job had to be done properly.

Meanwhile, Raynham got away at 7.40 a.m.

As soon as the magneto was back on the engine, Hawker got up again. There was no time for testing the repair beforehand; not even a trial spin. With thirty-two gallons of Shell petrol and seven gallons of Castrol oil filling the tanks, he taxied away from the sheds and took off. The time was 9.17 a.m. He rapidly ascended to 400 feet, far above Raynham. The engine spluttered and coughed for a few minutes as it warmed up, but it soon settled into a steady hum.

By mid-morning, any remaining wisps of fog over the aerodrome had completely cleared. After three hours making circuit after circuit, Raynham was getting bored. He throttled his engine right down until it was barely turning the propeller enough to stay aloft. Then, with the tail of the stocky biplane hanging down, he came in low over the spectators standing at the doors of the sheds. At times, he was just thirty feet above their heads. Then he started performing tricks, darting around over the huts, doubling back on himself, and flying round the circuit the wrong way. He was making the Avro team nervous. They dragged a hut door out into the open and laid it flat on the ground, before painting "Fly higher" on it in huge, whitewashed letters. Raynham paid it little attention.

Meanwhile, Hawker, at his 400-foot altitude, had only victory in mind. He flew round carefully and without flourish. A sandwich, some chocolate, and a flask of cocoa were all that broke the monotony. The time slowly crept by as he kept one eye on Raynham and the other on the clock.

By one o'clock in the afternoon, a rumour was spreading among the mechanics that Raynham's engine was about to run out of oil. Puffs of oily smoke, prominent at the start of the flight, were no longer being emitted. It was a telltale sign that levels were dangerously low. Worse still, the pressurized fuel tank feeding the Avro's engine had sprung a leak, spraying a constant fine mist of petrol directly into Raynham's face. But the taste of gasoline and the sting in the eyes was all part of the delight of flying. He flew on.

By two o'clock, nobody at the sheds could keep up the pretence of working. All eyes were trained skyward; all ears listened for the rhythmical *pop-pop* of Raynham's engine. By three o'clock, a tiring Raynham knew that his oil had finally run out, as his engine speed was dropping alarmingly. At ten minutes after three, Raynham figured the engine was about to seize. He brought the aeroplane safely in to land. He had been airborne for seven hours, thirty-one minutes, and thirty seconds. It was Britain's longest ever flight. He had broken the record.

Hawker was still up, but he was ninety minutes behind Raynham. He thought he had plenty of fuel and oil to keep going, but he throttled the engine down anyway, just in case. Soon after 4 p.m., Tom Sopwith turned up at the sheds. He climbed into another biplane, took off, and climbed rapidly to reach Hawker, waving and shouting encouragement as he flew alongside. After a few minutes he left Hawker and returned to the ground.

By half past four, the atmosphere at the aerodrome was shifting. That morning, nobody thought that the newly qualified Hawker, with his magneto trouble and his late start, could beat the experienced Raynham. But now, everybody was wondering whether the confident Aussie might be able to pull it off after all.

"Nine minutes . . . eight minutes . . . seven minutes . . ." The course timekeeper slowly counted down to the moment when Hawker would reach Raynham's newly set record. These few minutes felt like an eternity, but the countdown eventually ended. Soon after 4.48 p.m., a cheer went up. Hawker had beaten Raynham. He gunned the throttle and sped away. There was no need to worry about running out of fuel and oil any more. But he wanted to stay aloft for a while longer. Somebody else might try to beat him in the seven days that remained until the Michelin contest closed, and he wanted to deny them an easy ride. He gave himself a new target. Competition attempts had to be completed by one hour after sunset. On October 24 that meant 5.48 p.m.

On Hawker flew.

By 5.20 p.m., darkness had fallen. It was hard to navigate. The only landmark Hawker could see was the dim lighting of the Blue Bird restaurant amid the sheds of the flying village. He could no longer see his cockpit clock, so he had to estimate the minutes that passed. At half past five he opened the throttle full and launched himself to an altitude of 1,400 feet. Flames shot out of the red-hot exhaust of the engine. He wanted to get to the far end of the circuit quickly so that he could come in to land. He had tempted fate long enough.

As the minutes ticked away before the deadline, Hawker completed his final circuit of the aerodrome. Then he straightened up, set the throttle to 50 per cent, and came down in a perfect straight line. Mechanics lit gasoline fires to illuminate the ground as he approached. His landing was as soft as butter. He dropped down onto the grass at 5.41 p.m., just seven minutes before the competition time limit was reached. He had been in the air for eight hours and twenty-three minutes.

The bronze trophy Hawker later received depicted an aviator climbing onto the back of the winged horse of Pegasus, which struggled to leap into the sky. It held this inscription: *The Man, by the Aid of his Aeroplane, overcomes the Attraction of the Earth.* It was a new record—the second of the day—in a world that was changing fast.

CHAPTER 2

ENGLAND IS NO LONGER AN ISLAND

ALFRED HARMSWORTH MIGHT NOT HAVE FLOWN AEROPLANES himself, but as a cheerleader for aviation there was no louder voice. He was born in 1865 in a handsome villa by the River Liffey in Chapelizod, a picturesque village for the well-to-do and upwardly mobile in the western suburbs of Dublin. His parents wore the trappings of gentility, but the family lived a straitened existence. Alfred's father taught at a British military school in a park opposite the Harmsworth home. Despite the ever-present threat of Fenian reprisals, it should have been a comfortable childhood. But the family income was drained, and opportunities to rise in Dublin society were stifled, by the father's drinking. He was an alcoholic, usually filled with remorse but committed nonetheless. When Alfred was not quite two years old, the family sold up and moved to London. His father, who considered himself a gentleman, was retraining as a barrister. But life in London was no easier, and they still struggled to get by. In silent desperation the family headed downward, not up.

Alfred was a quiet child who kept himself distant from those around him. Aloofness from his immediate surroundings was matched by a growing fascination with the wider world, and the part he might play in shaping it. As he got older, he took to newspapers and magazines more than books. A neighbour remembered seeing him settled in a chair, reading. He was about eight years old at the time. "He seemed to be in

A teenaged Alfred Harmsworth (on the right) on a cycling excursion.

a world of his own, of which he was chief and king and sole possessor," she recalled.

Harmsworth performed well at school, but not exceptionally. He was becoming smart in a different way. His mother used to caution him, "Those who ask shan't have, and those who don't ask won't get." Young Alfred would reportedly reply, "Yes, Mother, but I take." He came to be disgusted by his father's drunkenness. He opted instead for the discipline of a self-controlled life. He grew strong and athletic, and took up cycling, which was a perilous activity on the high-wheel bicycles of the day. He matured into a confident teenager who was "extraordinarily attractive," as one school friend recalled. "He had golden hair and blue eyes," said the friend, "and carried himself in a commanding way. People often turned to look at him in the street."

At the age of fifteen, Harmsworth founded a magazine for his school, Henley House. He wrote short, breezy articles, jokes, and bad poetry, and he published readers' questions. He sold advertisement space, designed eye-catching headlines, and set the type himself in a local compositor's shop so that the articles seemed to jump from the page to the reader's eyes. He loved the power that journalism gave him over his peers. He decided what they should talk about. He controlled their minds.

At sixteen, having caused a scandal by getting the family's seventeen-year-old maidservant pregnant, Harmsworth left school, moved out of home, and became a freelance journalist. For five years, he wrote for anyone who would take his copy, whether specialist magazines or Fleet Street newspapers, and it didn't seem to matter whether he knew what he was talking about or not. He worked out what would sell and he stuck to it. He got a steady stream of commissions. But there was no real money to be made like this, and he barely scratched a living. He knew that the businessmen of London would only take a jobbing reporter seriously if he dressed smartly, but he couldn't afford the long frock coat that was the established uniform of the day. So, he bought one with his roommate and they operated a timeshare agreement on it. To save on bus fares, he walked around London and handed in his copy on foot.

His first few years of adulthood could have been a deadening experience, but he learned quickly. By 1886, at the age of twenty, he had secured the editorship of the *Bicycling News* in Coventry, home of Britain's cycle industry. The journal's circulation was dropping through the floor, but Harmsworth brought to it all the innovative ideas he had been developing since his school days. He shortened its articles. He introduced reader correspondence. He employed a female reporter so that the paper could treat the bicycle as an emancipatory revolution, not just a recreational pastime for macho teenage boys, as he had once been. His innovations were a resounding success, and the paper's fortunes were transformed. It was the first step in Harmsworth's rapid rise through the ranks of Britain's popular journalism.

Early in his career, Harmsworth spotted not so much a gap in the newspaper market but a growing chasm. A revolution had swept Britain in recent decades. Primary education had become universal and free, which meant that the great mass of the population was now literate. Thanks to changes in the political franchise, more and more of that population had the right to vote, which meant they could influence the very direction of the country. This paralleled a rapid growth in business, commerce, industry, government bureaucracy, science, teaching, and the professions. There was now a burgeoning lower-middle-class population of clerks and middle managers, the sort of workers immortalized as Charles Pooter and his friends in the 1892 work *The Diary of a Nobody*. This army of Pooters would make the long omnibus ride to work each morning from their sleepy suburbs to the crowded cities, and they'd need something to read on the way. But if they wanted the day's news, there wasn't really anything suitable. The existing roster of daily papers—the likes of *The Times*, the *Daily Telegraph*, the *Morning Post*, the *Standard*, and the *Daily Chronicle*—tended to be stuffy and complacent. Their fare comprised pompous leader columns, undigested reports of lengthy proceedings in Parliament and the police courts, insider society gossip, and the arcane movements of the money markets. There was nothing short, pithy, and straightforwardly written. There were no *stories*. There was no sense that the reader mattered.

"The Board Schools are turning out hundreds of thousands of boys and girls annually who are anxious to read," Harmsworth once told a friend. "They do not care for the ordinary newspaper. They have no interest in society, but they will read anything which is simple and is sufficiently interesting."

The older journalists on the establishment dailies never really got this. "Their standard of importance," explained a longstanding Harmsworth insider, "is set by the chiefs of political parties, Foreign Office and the Treasury; by the famous clubs (Reform, Carlton, Athenaeum); by the great country houses, the country rectories; by the Universities, by Bench and Bar. Now the standard is to be set by the mass of people."

What Harmsworth worked out wasn't so much what the public wanted, though he seemed to care more about that than did many of his rivals. His most powerful insight was that he could use newspapers to mould public opinion: not to give people what they wanted, but to tell them what they wanted. Harmsworth was just thirty years old when he founded the *Daily Mail* in 1896. It was a risk, but it quickly paid off, because the newspaper was a runaway success. He later bought and transformed *The Times* and *The Observer*, and he established the *Daily Mirror* in 1903. With his school magazine, Harmsworth dictated what his classmates would talk about. Now, it was the national conversation that he was controlling.

Harmsworth once explained, "Every day, there is an event which ought to be the outstanding feature of the news column. The clever news editor puts his finger upon that; the other man misses it altogether." He called them talking points. "Talking-points every day! Every day our pebble must be thrown into the pond," he would urge the older journalists on his staff. "You old stagers thought that 'news' was what you got sent in to you, the routine stuff, the ordinary. I tell you that the only 'news' worth the name is what you send out and gather." A Harmsworth editor once said, "The mass of people have no tastes of their own; they will adopt any that fall in their way. Give them a great deal to read about any topic within their comprehension: they will think they are getting what they want, will ask for more."

Harmsworth wasn't just an editorial innovator. He was obsessed with the technological innovations of the modern age as well. He had grown up an industrious devotee of modernity. His youthful bicycling obsession transformed into his interest in the earliest motor cars, which led him to the idea of a motor racing track at Brooklands. A telegraph cable linked the *Daily Mail*'s London office directly to New York via Ireland and Newfoundland. His newspapers were outfitted with linotype printing presses, automatic cutting and folding machines, and mimeograph duplicators. In the *Mail*'s first issue, he had boasted of "remarkable new inventions . . . the latest English and American construction."

He had a fastidious early-morning ritual whereby he reviewed the first editions of his papers each day. "In his bed, with a secretary sitting beside it," an insider observed, "he turns over the pages, scowls at them, utters cries of exultation, pounces upon ideas, pours out instructions." He would seize on what he felt landed well, critique what was missing the mark, and dictate how his team of editors, reporters, and writers could improve the next issue. "Get me the News Editor," he would bark at his secretary. "What? Won't be up! Why not? I'm up, or at any rate I'm at work. Get him. Pull him out of bed if necessary." There was always a telephone within reach. "I never think of him without thinking of telephones," a newspaper colleague observed. "He had them everywhere—at his bedside, at his fireside—one or two in every room. I dare say he had one in his bathroom."

On December 17, 1903, the American brothers Orville and Wilbur Wright made the world's first aeroplane flights. The longest of the session that day lasted just fifty-nine seconds and covered half a mile, and hardly anybody saw it. The brothers carried out their experiments in secret over remote sand dunes near Kitty Hawk on the North Carolina coast. It wasn't until November 1906, when the Brazilian aviator Alberto Santos-Dumont made a series of aeroplane flights in Paris, that the public started to catch on. They were the first flights in Europe, and to those with an interest in European geopolitics, they presaged an important shift. The technology-obsessed Harmsworth realized he had found another talking point for his *Daily Mail*. He comprehended, earlier than most, that aeroplanes would play a decisive military role. After seeing reports of Santos-Dumont's achievements, Harmsworth warned his news editor, "England is no longer an Island."

What was in his mind was a German invasion. He believed that war was coming.

GERMANS SEEMED TO BE EVERYWHERE IN BRITAIN IN THE LAST decades of the nineteenth century and the first years of the twentieth. Tens of thousands had migrated to London and the large industrial cities like Manchester and Glasgow in search of jobs, or business opportunities, or just a more liberal society. They worked in the service industries and ran small businesses: tailors, teachers, bakers, butchers, hairdressers. One in every ten waiters in London was German. Charlotte Street, a thoroughfare in the capital's West End where many Germans lived and where one in three businesses was German, had earned the nickname "Charlottenstrasse." Some Germans rose to the top of Britain's industrial and financial institutions. Others brought their musical talents and performed at every level, from the street to the finest concert halls. Many more—the very poorest—sweated over vats of boiling sugar in the refineries of London's East End.

Germany bought 20 per cent of its food and raw materials from Britain and its empire. In return, Britain bought almost 15 per cent of Germany's exports. Most of Germany's merchant ships were insured by British underwriters. Much of its banking was transacted in the City of London—often by the German financiers working there. Germans in Britain had their own newspapers, their own Lutheran churches, and their own clubs. They took British wives and husbands. Young German governesses tutored the children of Britain's upper classes. Queen Victoria's mother was German, her husband was German, and she was the grandmother of Germany's emperor, Kaiser Wilhelm II. Her son Edward, who acceded to the British throne in 1901, was Wilhelm's uncle.

Britain and Germany seemed economically and socially inseparable, which was why it was so hard for people to imagine that a serious conflict might arise.

Germany unified as a nation in 1871. By the late 1880s, its status as an industrial, military, and imperial power was ascending rapidly. Its close economic and social ties with Britain masked the belligerent Kaiser Wilhelm's desire to compete with, and surpass, his rival in Europe. Alfred Harmsworth, a British imperialist to his core and a keen analyst of European politics, saw only enmity behind the veil of friendship. He had been watching German hostility towards Britain rising for years. He believed that the kaiser had Britain's empire in his acquisitive sights, and he considered it his patriotic duty to use the power of his press to sound the alarm. A close friend recalled that, by the time he founded the *Daily Mail* in 1896, Harmsworth "already knew that Germany had begun to make ready for Armageddon, and that sooner or later she would strike." An industry insider observed that "the 'German Menace' provided him with a dominating talking point for his *Daily Mail*—a talking point which was to go on almost non-stop for the next eighteen years."

Harmsworth seemed to have an instinctive awareness that aeroplanes would play a critical role in any emerging European power struggle. But the British public and its government would have to be made to care about them. This was what *Daily Mail* talking points were designed for. In Harmsworth's editorial hands, the aeroplane would be presented as an existential threat to the island nation of Britain, hitherto protected by its powerful navy. "This news," Harmsworth said to his editor as they discussed Santos-Dumont's flights in 1906, "means no more sleeping safely behind those 'wooden walls of old England' with the Channel our safety moat. It means the aerial chariots of a foe descending on British soil if war comes."

But the aeroplane could bring salvation as well. One of Harmsworth's editors later reflected that, "as he had made up his mind that aerial supremacy would be a deciding factor in any war with Germany, he brought his tremendous energy to what he came to accept as a mission for national security—to see that Britain should be wakened up to capture leadership in the air, whether for peace or war."

A few days after Santos-Dumont's flights in Paris, the *Daily Mail* announced a prize of £10,000—the equivalent of £1,000,000 today—to the first aviator to fly from London to Manchester. As a confidant later put it, Harmsworth wanted to "stimulate British airmanship and British aeroplane construction and to make the British public air-minded."

DESPITE HARMSWORTH'S FINANCIAL INCENTIVE, THE FIRST flights in Britain did not take place until two years later, in 1908. Alliott Verdon-Roe, the man who would later name his company Avro and give Fred Raynham a job, is said to have made short, unofficial hops in his biplane at Brooklands early in the year, long before the flying village was built. It was the American showman Samuel Cody, who took to the skies above the Hampshire town of Farnborough on October 16, who made the first official flight. But on the Continent, aviation by then was moving into the mainstream. Over a period of two weeks in August 1908, at a racecourse near Le Mans, in France, Wilbur Wright performed a series of nine aeroplane flights in front of an audience of hundreds. The first lasted 107 seconds; the last totalled 8 minutes and 13 seconds. One experimenter who witnessed them was the French aviator Louis Blériot, who exclaimed that "a new era in mechanical flight has begun. I am not sufficiently calm after the event thoroughly to express my opinion. My view can best be expressed in these words—it is marvellous!"

Wilbur Wright stayed in Le Mans for the rest of the year, making flight after confident flight, each to an amazed audience. During one ascent, he carried a journalist from *Le Figaro*. Afterward, the reporter wrote, "I have known to-day a magnificent intoxication. I have learned how it feels to be a bird. I have flown. Yes, I have flown! I am still astonished at it; still deeply moved. For nearly an hour I have lived that daring dream vainly pursued through all the ages by audacious man."

In the new year, Wilbur Wright relocated south to the city of Pau, near France's border with Spain, where he was joined by his brother Orville. Among countless visitors who met the aviators that spring was Alfred Harmsworth. By now, the newspaper magnate was forty-three years old. Success had brought him money and power. His once-beautiful features had coarsened and hardened, and his formerly athletic frame had filled out with the excesses of fine living. But he moved through life with his barrel chest puffed out more confidently than ever before. The child who boasted that he took what he wanted instead of asking for it was now a married man who had mistresses, and affairs, and illegitimate children. In 1905, he had been raised to the peerage as Lord Northcliffe. He lived at the top of British society; his counsel was sought daily by government ministers and business leaders. The world of politics was the stage on which he now acted, but always as a fiercely critical outsider. He was a patriot, but to the British people, not its government.

At Pau, playing the role of everyman, Lord Northcliffe wore a cloth cap, rather than the bowler hat or topper that signified the elevated echelons of Britain's class hierarchy. When the Wrights needed extra hands to haul launching equipment in readiness for a demonstration flight, Northcliffe gamely took up a spot on the rope alongside former Prime Minister Arthur Balfour—also in a flat cap—and a clutch of mechanics

Alfred Harmsworth, Lord Northcliffe (in the fur coat), former Prime Minister Arthur Balfour (to his right), and Orville Wright (in the bowler hat) helping set up for a demonstration flight at Pau in 1909.

and assorted helpers. Northcliffe's man-of-the-people act was compromised by the luxurious fur coat he wore, which was so long its thick hems dragged in the French mud and almost got caught under the heels of the next man pulling the rope. But Northcliffe, with a fat cigar clamped in his grinning mouth, didn't care about the incongruence of his appearance. Perhaps he remembered the time he couldn't afford a whole frock coat and had to go halves with his friend. He had worked hard to elevate himself since those days. Now he was a big man dragging the world into the modern age.

Lord Northcliffe was impressed by what the Wright brothers achieved, but he noted with exasperation that the British government failed to send a single observer to witness these world-changing flights. Back at his desk in London, he wrote to Britain's secretary of state for war, Richard

Haldane, to express his frustration. A government official replied that Britain's naval and military experts considered aeroplanes "a very long way off being the slightest practical use in war."

Northcliffe had previously lamented to a friend that the British government "regarded aviation as a silly fad of the *Daily Mail*." He was a mere newspaperman, not a government minister or a military strategist like Haldane and his officials. But if he could see how important aviation was about to become, why couldn't they? Somebody had to wake the establishment authorities from their complacency. He figured it would have to be him. In early October 1908, he announced a prize for the first person to fly an aeroplane across the English Channel. Whoever made such a flight would prove his prediction that England's wall of naval protection could now simply be stepped over. The military potential of the aeroplane would be evident to all.

He didn't have to wait long for the challenge to be met. On July 25, 1909, the Frenchman Louis Blériot piloted the first aeroplane to cross from continental Europe to England. His aircraft was a delicate little monoplane; a journalist said it looked "rather like the skeleton of a huge blue-bottle fly." The route took Blériot from Calais to Dover across the narrowest part of the Channel, which is a mere twenty-three miles wide at that point. The moments after he cut his engine to land almost took him by surprise. "In two or three seconds I am safe upon your shore," he recalled the following day. "Soldiers in khaki run up, and a policeman. Two of my compatriots are on the spot. They kiss my cheeks. The conclusion of my flight overwhelms me."

It had taken Blériot just thirty-six minutes to win £1,000 from the *Daily Mail*—nearly £100,000 today—and to change the politics of Europe. Northcliffe's private secretary recalled what he had said to her that day: "Do you realize it is the first time an entry has been made otherwise than by ship?" The next day, the *Daily Mail* wrote, "The British people have hitherto dwelt secure in their islands because they have attained at the price of terrible struggles and of immense sacrifices the supremacy of the sea. But locomotion is now being transferred to an element where

Louis Blériot flying towards
Dover's chalk cliffs, July 25, 1909.

Dreadnoughts are useless and sea power no shield against attack. As the potentialities of the aeroplane have been proved, we must take energetic steps to develop a navy of the air."

The following month, a week-long international flying meeting took place at Reims, to the north-east of Paris. It was the world's first. North-cliffe attended. This time, unlike at Le Mans and Pau, so did representatives of the British government. One was David Lloyd George, chancellor of the exchequer and a future wartime prime minister, who said, "Flying machines are no longer toys and dreams; they are an established fact. The possibilities of this new system of locomotion are infinite. I feel as a Britisher rather ashamed that we are so completely out of it."

Through the pages of his newspapers, Northcliffe remained an outspoken critic of the government's air defence measures as time went on and the threat of war with Germany grew more insistent. In 1912, the Royal Flying Corps, with army and navy wings, was formed. Two years later, the naval wing split off to form the Royal Naval Air Service. At last, aviation was starting to get built into the structure of Britain's armed forces. But progress was slow, and aviation still had plenty of critics among military leaders.

Northcliffe announced a string of further prizes in the *Daily Mail*. There was one for another cross-Channel flight. There were circuit races of Britain. There were prizes for cross-country races and aerial derbies. The first derby, held in 1912, was won by Tom Sopwith, flying a Blériot machine. By then, aeroplanes were being flown in places around the world, by aviators of all classes and backgrounds. In aviation, it didn't matter who you were, or where you had come from, or how much money you had, or where you stood in society's order. All that mattered was that you were obsessed with flight. Aviators talked of nothing else while they were awake. When they slept, aeroplanes filled their dreams.

CHAPTER 3

ONE OF THE SOUNDEST PILOTS IN
THE COUNTRY

THERE WAS ONE YOUNG MECHANIC AT BROOKLANDS PAYING PAR-
ticularly close attention as Harry Hawker and Fred Raynham competed for the Michelin endurance prize in October 1912. He was a
friend of theirs. In time, the three would become closer than they
could have imagined, in locations and circumstances that would have
the whole world looking on. At that moment, the mechanic was less
than a month from taking his own flying tests. As he watched Hawker
and Raynham fly overhead, he was scrutinizing their technique to help
perfect his own.

Jack Alcock was a tall nineteen-year-old, like Raynham, but stock-
ier, with a round face and softly smiling eyes. He had strawberry-
blond hair, which he pushed back over his head, and ruddy cheeks,
which would blush a hectic brick red whenever he was embarrassed.
His first name was John, but people invariably called him Jack, and
he was the sort of teen that everybody got on with. He was affable and
kind. Like Hawker, he showed a spirited sense of humour. But Alcock's
cheery jokes masked an inherent shyness that contrasted with Hawker's
spikier character.

Alcock's thick Lancashire accent had been nurtured in the streets of

Jack Alcock in 1912, aged twenty.

south Manchester, where he had been born in 1892 to a coachman father, John, and a mother, Mary, who cleaned rooms and served drinks at a local public house. Jack's family moved around to follow work. Mostly, they stayed in the Manchester area, but when Jack was about seven, they decamped for a while to St. Anne's-on-the-Sea, a coastal resort town forty miles to the north-west.

They moved back to Manchester in 1905, when Jack was twelve. At school, and at the local Sunday school he attended each week, Jack was a diligent pupil. When studying was over, he earned pocket money as a delivery boy for his aunt's husband, who ran a butcher's shop. Families in the area got to know well the smiling and cheeky teenager who called on them to hand over their orders. At his own home, his mother taught him to cook. He was happy in the kitchen. His younger brother Albert would later liken Jack to a famed English cookery writer. "He could make a superb parkin, roast a turkey, fry a steak to tender perfection and produce an apple pie with the flair of Mrs. Beeton," Albert recalled. Jack also made a mean treacle toffee, though he rarely got to eat much of it himself. Another brother, Edward, observed, "I can well remember my sister, my two younger brothers and myself sitting in a circle round him like a pack of wolves." Their mother would reportedly say, "Jack would make a very capable husband for some lucky girl."

Jack Alcock turned sixteen in November 1908, and like the rest of his classmates, he needed a full-time job. Three in every four Britons at that time lived in towns or cities, but life there could be precarious. There were plenty of options for work: neighbours along Jack's cramped red-brick terrace around this time included a boot repairer, a warehouseman, a wool dyer, a couple of laundresses, some domestic servants, and two teenage sisters who worked in the local chain-making factory. But working life for most was regimented and repetitive, and jobs could become all-consuming. Hours were long, which meant there was little time for outside interests. Wages for men tended to be low, and women's wages were far lower. Something like one in every three workers lived in poverty. Those families who normally kept themselves above the breadline would nonetheless most likely endure periods of temporary poverty. There was rarely much surplus to put by for a rainy day. Working families could be one illness, one accident, or one bout of unemployment away from destitution. The state still held to the Victorian belief in self-help. There were always the workhouses if things got too desperate—they were still doing a roaring business in 1908. What little income there might have been to spare would probably be paid into funeral insurance.

But there was a sense of hope growing in places like Manchester as well. Demands for increased suffrage were gathering momentum. Authority was being challenged. People like Jack Alcock, as well as his friends, family, and the people who lived along their close-knit street, now had representation in Parliament through Labour, a party for the millhands and the miners, the factory workers and the railwaymen. The other parties were being forced to consider the workers as well. In a 1908 parliamentary debate on working hours, Winston Churchill, then a Liberal, spoke out for working people. "They are not content that their lives should remain mere alternatives between the bed and the factory,"

Churchill said. "They demand time to look about them, time to see their homes by daylight, to see their children, time to think and read and cultivate their gardens—time, in short, to live." The stifling strictures of class remained, but people could sense change in the air. The passing of Queen Victoria in 1901 felt like the end of an old age. The new age was modern.

The growth of new industries and technologies offered opportunities to those who wanted change: opportunities to step off the predestined path. For a sixteen-year-old boy looking for work in 1908, Manchester could have been a depressing place. It could also be a paradise.

A FRIEND OF JACK ALCOCK'S FATHER WAS THE WORKS MANAGER of the Empress Motor Company, a motor vehicle business that was a half-hour walk from the Alcock family home. He heard that Jack was in the job market and offered him an apprenticeship at Empress. Jack readily accepted. He was glad to be working with his hands. A fellow Empress employee once recalled that Alcock was "for ever tinkering about with machinery. You couldn't keep him away from it. There never was such a youngster. He was keen as mustard on anything and everything to do with wheels."

Alcock's new apprenticeship meant he could get to grips with the internal combustion engines that were a speciality of the firm. Motor cars and motorcycles became his life. But there would be more to his posting at Empress than everyday garage activity. The firm was run by Charles Fletcher, known as Charlie, a man who harboured an ambition to be the first person in Manchester to fly. In 1909, Fletcher formed a small alliance with other motoring pioneers in the city, including Billy Turner, who sold Henry Royce his first motor car, and William Arnold, who ran a coachbuilding firm. Together, the syndicate resolved to build their own aeroplane from scratch, engine and all. They invited Jack Alcock to join them.

That summer, Charlie Fletcher and his crew of engineers, apprentices, and collaborators began constructing a monoplane, and they made fast progress. By September 1909, most of the work on the bamboo-framed aircraft had been completed and the group was finishing the engine. Alcock—a big, strong lad, others on Fletcher's team recalled—was given the task of swinging the engine's propeller to start it for tests. Nobody else could reach it on its tall test-frame.

Come October, the monoplane was ready for trials. The site chosen by the group was Manchester's Heaton Park, a 650-acre municipal open space to the north of the city that offered the expansive sloping ground needed for long taxiing runs. Fletcher had also been given the use of a large shed within the park to house the aeroplane between trials.

The early morning of Wednesday, 13 October was accompanied by customary Manchester rain, but it didn't dampen Alcock's spirits as he helped assemble the aircraft and make minor alterations. Before long, the preparatory work was completed, and the trial was ready to begin. The engine refused to start.

A week later the team was out again, and this time they had more success. Charlie Fletcher seated himself on the wicker armchair strapped to the wooden frame. This time the engine started. After a couple of hundred-yard rolling runs, the aircraft finally made it into the air. It was a short flight: Fletcher only managed a hop of about twenty-five yards before a gust of wind brought the machine back down with a thump in an undignified landing. But at least he made it off the ground. A small crowd of local children cheered loudly in appreciation. They were beside themselves at seeing an aeroplane take off in front of their disbelieving eyes.

What a thrill it must have been to be among the first humans to fly. How many children for millennia past looked into the open skies and dreamed of joining the birds there, ascending and soaring and swooping through the air without obstacle? How many, in moments of sorrow or weakness, pored over the Old Testament words of Isaiah: "They that wait upon the Lord shall renew their strength; they shall mount up with wings as eagles." How many wished they could shake off Earth's shackles, even for a few fleeting moments, and be free? To be among the first to fly was to be among the gods.

In December, after some modifications, more trials of Fletcher's monoplane followed. Early in the new year of 1910, after the machine had been damaged in another crash landing, Fletcher had a newer, improved version to try out. Later that year, still with Alcock involved and having moved the trial ground to Manchester Racecourse, Fletcher's syndicate built a biplane. More short hops followed. There were no sustained flights, but they seemed to be making progress.

By now, Alcock was hooked. He was unable to confine his new-found obsession to working hours. At home, he experimented with silk and bamboo hot-air balloons and small model aeroplanes, eager to understand the principles and practicalities of flight. And he read books—lots of them. A journalist later wrote, "There was never a book written on the subject of mechanics—text books, manuals, or inch-thick volumes stodgy with stark facts and diagrams—that did not ultimately find its way into that young man's hands, and were one and all devoured by the youthful mind with the keenest relish." Alcock's family, reportedly, looked on in amusement, but soon they would discover just how seriously young Jack was taking the new world of aviation.

IN JANUARY 1911, EMPRESS SENT ONE OF ITS AEROPLANE ENGINES down to Brooklands, where it had been ordered by the Avro flying school. Charlie Fletcher asked Alcock, who had recently turned eighteen, to go as the mechanic-in-charge to set it up. Alcock went by motorcycle. When he arrived, he rode across the airfield and stopped at shed number 15, where the Avro school was based. Howard Pixton, Avro's twenty-five-year-old flying instructor, was there with the engine. He and Alcock got chatting. All around them, the flying village was alive with activity. For Alcock, the sense of opportunity on display at Brooklands that day was exhilarating. At one point during the conversation, he asked Pixton if he knew of any jobs going at the airfield. Pixton walked him to the neighbouring shed and introduced him to Maurice Ducrocq, Hilda Hewlett's first pupil. Ducrocq offered Alcock a job as his chief mechanic. Within days, Alcock was sitting by Ducrocq's side, flying high over the Brooklands track.

The affable lad from Manchester adored his new position, and soon, everybody in the flying village got to know him. His enthusiasm for the work was unquenchable, and so was his spirit. He got hold of a wind-up gramophone and a stack of records, set them up at the back of Ducrocq's shed, inserted the loudest needle he could find, and spent his days belting out the latest popular tunes while he worked on Ducrocq's aeroplane. "At Weybridge in 1911, everyone," his brother Albert would recall, "knew when he was in the hangar."

Alcock soon wrote to his Empress colleagues in Manchester, asking them to send on his final wages. He told them he was going to stay in London for a while. His adolescent life in Manchester was now behind him. He had spent it, in the words of one later observer, "sweating all day amid the oil and the grime and the stench of the stifling garage." It hadn't been so bad: he'd loved the work and the opportunities that Charlie Fletcher had given him, and he'd made good friends there. But he always wanted to be challenged. Brooklands, and the open sky above it, was his home now. He would still be sweating all day, but for a higher purpose: the freedom and joy that flying could bring.

There was always a price to pay for freedom, though. Jack Alcock soon learned that first-hand.

ON AUGUST 1, 1911, A BROOKLANDS PUPIL, GERALD NAPIER, crashed his new single-seat racing biplane. It was just two weeks since he had qualified as a pilot. There had been countless crashes at Brooklands, but each time hitherto the pilot had been able to walk away. This time, there was no lucky escape. Napier was the first aviation fatality at Brooklands, and only the fourth in Britain, although there would soon be many more.

Napier, like Jack Alcock, was nineteen years old. But their ages were all the two young men had in common. Alcock came from a working-class Manchester family. Napier, the son of an army major killed fighting in the Mahdist War, had been educated at Eton. He quickly earned a reputation among the Brooklands community for his arrogance and recklessness as he learned to fly that summer. On the morning of the crash, Howard Pixton, Napier's former instructor, offered to take the newly delivered machine up first. He wanted to advise Napier on how the biplane's handling differed from that of the school machine. The racer was smaller and much faster; its controls would need a sensitive hand. But Napier, full of conceit and impatience, refused the offer. After all, he was now a qualified flyer. He would do what he liked.

"Like a baby with a new rattle, he wanted to use it at once?" the coroner asked Pixton at the inquest, two days later.

"Yes," Pixton replied.

One eyewitness recalled, "The moment he started out on his new machine everyone knew there was going to be a smash, and several men went back into their sheds rather than stand and watch him, hoping that when the smash came it would let him off lightly."

During his first three flights, it was obvious that Napier was struggling to command the machine. He kept getting his controls mixed up and was relying on the power of the engine to get him out of trouble. It was on his fourth flight that the crash happened. Napier decided to take his close friend, Edwin Laurie, as a passenger. There was no passenger seat in the little racer, so Laurie had to sit astride the petrol tank, behind Napier. This raised the centre of gravity and changed the handling of the powerful aeroplane still further from what Napier was used to. After a couple of circuits, it looked like he might get away with it. Then a gust of wind caught him as he was banking too tightly at low altitude. The aircraft stalled and side-slipped downward before starting to spiral. An experienced flyer could have recovered. The conceited Napier, in his panic, opened the throttle full, and drove the machine nose first into the ground.

Jack Alcock saw the whole thing. He ran to help. Edwin Laurie, from his precarious position behind Napier's seat, had been thrown clear of the aircraft and knocked unconscious. Napier wasn't so lucky. When Alcock reached the scene, he pulled aside the tangle of wreckage. When the aircraft hit the ground, a thick wooden strut had snapped off. It had then been driven, like a stake, through Napier's mouth. It emerged from the side of his throat below the ear. Its impact had snapped Napier's skull backward, breaking his spine and pinning his head to the hard, grass-covered ground. Alcock pulled the strut out of Napier's shattered jaw and lifted his limp body from the debris. Long before a local doctor arrived, Napier was dead.

His distraught mother came straight to Brooklands to visit the site of the crash and sit with Gerald's mangled body. She had already lost her husband to a violent death, and now her son had been killed. As for Edwin Laurie, it took two days before anybody could find the courage to tell him his best friend was dead, though he kept pleading for news. They couldn't bear to break his heart.

THOSE EARLY AEROPLANES WERE ETHEREAL, LIGHTWEIGHT CON-fections of lacquered fabric, wooden struts, and wire. They had fickle, underpowered engines that gulped fuel and oil and could only run for a few hours between rebuilds. But this fragility and unpredictability was what provided the thrill. Crashes were not just commonplace but a necessary part of the deal. The Brooklands pioneer John Neale, in 1910, said, "Every day's work ends in a smash, because, with an experimental machine, one simply goes on till a smash of some sort puts an end to the experiments for the day." Deaths, thankfully, were less common. But they, too, came to be accepted, as Warren Merriam, a Brooklands instructor who came to know Jack Alcock well, observed. "Gradually one or another would be taken from our midst," he said, "and we had to become hardened to the fact that in our calling nothing was certain. 'Here today and gone tomorrow' was the maxim we learned to accept, but, however deeply these losses may have grieved us, the fascination of our work was too great to deter us from any further sacrifices fate might demand."

One commentator tried to explain what motivated the world's first aviators. He claimed that "no cocaine, morphia or laudanum could pro-duce in twenty years that complete consciousness of aloofness from the world's cares, as one hour's flight in the air." Brooklands, along with a new aerodrome at Hendon in north London, had become home to anybody in Britain with an appetite for gasoline, adventure, and an urge to wager on gravity. Some, like Gerald Napier, lost the bet. But there were plenty who were prepared to accept the odds. Jack Alcock was one of them.

On November 16, 1912, Alcock passed his own pilot exams, though he had been flying unofficially for a while by then. It was ten days after his twentieth birthday and less than a month since his friends Harry Hawker and Fred Raynham had duked it out in the skies over Brooklands for the Michelin prize. Now Alcock could act officially as Maurice Ducrocq's pilot as well as his chief engineer. Proud of his new vocation, he had a set of calling cards printed. They simply read, "Jack Alcock, Brooklands, Weybridge. Aviator."

AFTER THE EXCITEMENT OF THE MICHELIN COMPETITION, LIFE AT Brooklands settled into an agreeable rhythm for the three young aviators, and for everyone else working there.

Harry Hawker continued to flourish at Sopwith and became increasingly involved with the aircraft manufacturing side of the business at Kingston. His new responsibilities sometimes inflamed the tension that had existed between him and his workmates since he first joined Sopwith. One acquaintance recalled, "They disliked it at the factory when he told them that their ideas of aircraft were out of date. He told them things about building for speed and endurance, building for more power, and they laughed at him." He came across, said the friend, like a "bombastic young Aussie." But they all grew to trust his judgement. The firm's draughtsman, Reginald Ashfield, said, "As soon as Hawker started to fly he had an angle we hadn't. He could tell where the shoe pinched. He was a damned annoying blighter, but he was *right!*"

Fred Raynham became much in demand as a freelance pilot carrying out test flights for manufacturers including Avro and Sopwith. He also worked for another company based at Brooklands: the firm founded in 1908 by Helmuth Martin and George Handasyde that became known as Martinsyde.

Jack Alcock entered his first competition just days after he qualified. It was a speed handicap in which he took first place, beating Tom Sopwith into second. In the months that followed, he competed in many more, quickly developing a reputation for his skill and endurance. In May 1913 Raynham and Alcock collaborated on an Avro project. They tested a seaplane which, with a wingspan of fifty feet, was the largest aircraft the firm had yet constructed. The following month, it was bought by the German government, which made a series of copies that would be used the following year by the country's navy.

Later that year, Maurice Ducrocq closed his flying school to concentrate on a sales business he had been running on the side. Alcock began test-flying for Louis Coatalen, chief engineer at the Sunbeam company. He was probably introduced to Coatalen by Billy Turner, the Manchester motor car dealer who worked with Alcock on aeroplane experiments four years earlier. It was said that Alcock's extensive testing of the prototype Sunbeam engine, fitted to a Maurice Farman biplane, meant he was probably spending more hours in the air than any other British pilot at the time. He came third in a race from London to Manchester and back—a remarkable achievement. A commentator later observed, "He was usually the first up in the morning and often the last down at night. He flew in weather which kept all the other pilots indoors, and he ranged far and wide over southern England. . . . the intense flying practice served to develop Mr. Alcock into one of the soundest pilots in the country."

Aviation had been developing rapidly since Alcock started experimenting on the grassy hills of Manchester in 1909. What had been a pastime for the brave or foolhardy was fast becoming a small industry. What had seemed like impossible dreams were starting to look like achievable prospects.

Then, in April 1913, Lord Northcliffe's *Daily Mail* put up yet another cash prize. This one, for another £10,000, was for a proposition so bold it garnered worldwide attention: the first flight across the Atlantic Ocean.

Jack Alcock and the Sunbeam-Farman aeroplane, 1914.

Most people thought such a feat was utterly impossible. Airframes were still slight, fragile affairs, and engines were underpowered and unreliable. The sceptics believed such experimental craft could never manage a flight as lengthy and arduous as the Big Hop. But a few adventurers thought differently.

By May 1914 two companies were in serious contention for the transatlantic prize. One was Glenn Curtiss in the United States. The other was Britain's Martinsyde company. Martinsyde's machine, a huge monoplane to be flown by pioneer aviator Gustav Hamel, would be powered by two Sunbeam engines—production versions of the prototype that Jack Alcock had been so assiduously testing at Brooklands. A route had been plotted from Newfoundland to Clifden, on the west coast of Ireland, and construction of the aeroplane itself was well underway. The attempt on the prize was planned for late July, and its promoters believed it held every promise of being successful. But on May 23, the twenty-four-year-old Hamel disappeared on a flight over the Channel. Neither his body nor his aircraft was ever found. It was a devastating blow for British aviation, and the end of Martinsyde's attempt on the Atlantic.

Before plans could be worked out with a replacement aviator, and with construction of the Curtiss machine delayed, world events overtook the competition for the *Daily Mail*'s transatlantic prize.

As of 11 p.m. on Tuesday, August 4, 1914, Britain and its empire were at war with Germany.

CHAPTER 4

THE GAME SEEMED GOOD

FOUR WEEKS AFTER WAR WAS DECLARED, TED BROWN, A TWENTY-eight-year-old engineer working in Manchester, enlisted as a private in the British Army. He later said, "Although of American parentage and possessing American citizenship, I had not the patience to wait for the entry into the war of the United States."

Brown was born in Glasgow in 1886 to an American couple, Arthur George Brown and Emma Whitten. Named Arthur Whitten Brown, he adopted the name Ted in childhood and kept it for the rest of his life. His mother was a Pittsburgh native. His father, originally from Schenectady, New York, had moved to Pittsburgh to work for the engineer Herman Westinghouse, and he'd been in Glasgow to set up a British manufacturing deal for a compact steam engine that he and Westinghouse had developed together. In 1892, having decided to leave Westinghouse employment and stay in Britain, Brown moved the family south to Bolton, then to the affluent Manchester suburb of Chorlton-cum-Hardy.

In Manchester, Brown ran a series of engineering, metalworking, and machining companies that specialized in motive power and in the alloys used in bearings and castings. A steady stream of patents in his name showed that engineering was second nature to him. Everything from engines to foundry machinery came under his inventive hand. The engineering businesses also took the family around the world, as Brown sought licensing deals and sourced new materials and products overseas. This gave Ted, the couple's young son and only child, an exotic upbringing, at least compared with most Manchester children. As well as trips around mainland Europe, the family shuttled back and forth across the Atlantic—on business, and, occasionally, to visit family back home.

Arthur's engineering obsession rubbed off on young Ted, who later recalled, "From my earliest days I had determined to follow in his footsteps." Once, when Ted was nine years old, his father set him to work on his own little engineering project. His job was to prepare shellac, a resin varnish used to bind the electrical coils in motors and transformers. He was also charged with splitting mica, a type of mineral, into the flat, thin sheets that were used to insulate electrical components. All morning, he would sit, dressed neatly in a sailor suit, carefully grinding the shellac in a coffee mill and working through his pile of mica. Once the job was completed to his exacting standards, he presented the results to his father, who paid him five shillings in wages for the job. Ted could have spent the money on toys, as most nine-year-old boys would have done. On a recent trip to a toy shop, a model train-set with a pair of lever-operated railway signals caught his eye. He also fancied a toy gun, and a seven-inch-long cannon that could be loaded with gunpowder to fire real shot. Instead, proud of having earned a wage doing valuable work for his father, the studious little engineer invested in his first camera and a set of developing chemicals.

After reaching his teens, Ted started collaborating with his father on inventions they patented together. Then, it was time to stand by himself. At sixteen, he took up an engineering apprenticeship at Manchester's British Westinghouse, a spinoff of the Pittsburgh corporation where his father had started out. Classes at Manchester's technical college came as part of the package. In 1906, after four years of diligent study, twenty-year-old Ted joined British Westinghouse as a qualified assistant engineer. The patent applications—now in his own name—continued to come.

The solitary little boy in the neat sailor suit quietly splitting mica had grown into a serious young man with a career wide open in front of him.

He grasped the opportunities that Westinghouse offered with vigour. He was stepping into a world that was in convulsion. New technologies poured out of factories and laboratories staffed by specialized engineers and professional scientists. There were new manufacturing techniques, new machine tools, new analytical processes, new business structures. Giant corporations like Westinghouse acted as crucibles for innovation. Then they scaled it up for industrial production. The four-decade period before the Great War saw the development of electric lighting and power, synthetic drugs like aspirin and heroin, a chemical industry, artificial fibres, radio, the telephone, cinematography, mechanical data processing, the machine gun, motor vehicles, and the aeroplane. Countless other innovations we take for granted today first emerged in that short, vivid period.

Over the next five years, Ted Brown came to specialize in electrical mining equipment. One of his earliest projects was to superintend the installation of a winding engine at a large coal mine in Pontypridd, Wales. Winding engines were the powerful machines that hoisted miners up and down the lengthy mine shafts, along with their tools, explosives, and material such as timber propping. They also hauled up freshly dug coal from the underground tunnels, bringing it to the surface where it could be processed. The new Pontypridd engine was Britain's first to be driven by electricity rather than steam, and Brown's success with its installation impressed his bosses. In early 1910, they sent him to South Africa for a year to install winding engines in the gold and diamond mines around Johannesburg.

It was tough work for the twenty-three-year-old. He had to labour hard each day, moving from mine to mine as he was needed. The conditions were far more challenging than he'd experienced in Pontypridd. At the Brakpan mine, twenty miles outside Johannesburg, he discovered that the heavy electrical machines shipped over from Manchester would have to be hauled for almost two miles on ox-carts over unmade tracks to reach the mine shaft. He also found that the mine didn't have a crane to hoist the powerful motors into place. All he could use was a makeshift gantry. Moreover, construction of the engine house itself had

Ted Brown's electrical installation underway in an uncompleted building at the Brakpan Mine, 1911.

been delayed. Brown had to install the machinery before the building was remotely finished. It was the rainy season in South Africa, and the complicated machines, with little protection, would be drenched. They were also exposed to the thick clouds of dust that blew across the region, coating every surface with a grinding paste that risked damaging the engines once they were put into commission.

Whenever he had a little time off from his heavy workload in the mines, Brown would help organize races for the Transvaal motoring scene, which was flourishing among the wealthy mine-owning and engineering elites there. In turn, this brought him into contact with aviation enthusiasts who were making South Africa's first flights at the time. He helped to organize the newly formed Aeronautical Society of South Africa, founded in Pretoria in March 1911. He also collaborated with Adolph Brunett, a French aviator working in Johannesburg, on the construction of a Henri Farman biplane, which first flew in May that year.

It was only a 150-yard hop, reaching a height of twenty feet from the ground, but it proved that the machine was airworthy, and it gave Brown his first practical experience of flying. "I had always longed to be in the air," he would later say.

But Brown's involvement with South African aviation and motor racing was to end before he could get deeply involved. In July 1911, the Westinghouse installations were nearing completion, and it was time for him to return to Manchester. Once he got back, he began to get papers published in professional journals and started giving lectures on the engineering circuit across Britain. Work trips representing Westinghouse in mainland Europe and America followed. Brown was growing in skill, experience, and confidence.

As the new year of 1914 dawned, he was back in the United States. Amid Christmas visits to family and friends, he was there for another stint at Westinghouse's Pittsburgh plant. He had just signed a fresh two-year contract with British Westinghouse. He had been promoted from an assistant engineer to a fully-fledged electrical engineer in his own right, and his salary had risen to £25 per month, which gave him a measure of confidence in his future. On this Pittsburgh visit, he was studying electrical components that were being taken up by railway companies and transit authorities around the world. Then, he returned to Manchester. The earnest and studious adolescent had matured into a widely travelled adult, as comfortable in Africa or America as in Manchester, and prepared to have a go at any mission.

When the war came, it was no different. He had a job to do, and he would apply his mind and his body to the task and give it his all. He could see a better future for the world and was ready to play his part in securing it. In 1914, he felt positive; perhaps he was even happy.

AFTER FOUR MONTHS OF ARMY TRAINING, BROWN RECEIVED A commission as a second lieutenant in the Manchester Regiment, followed by more training. In June 1915, he departed for the Western Front. He later recalled—tersely—that there, he "saw service in the trenches before

Ypres and on the Somme." More than fifty thousand British, Canadian, and Indian soldiers were killed, injured, or taken prisoner in the three months Brown served at the front.

There could, however, be found a better way out of the mud and gore of trench warfare. One young infantryman, Wilfred Blake, secured a transfer to the Royal Flying Corps, the aviation wing supporting army operations with reconnaissance and artillery spotting. Recalling his early experiences of the war, Blake described the "confused jumble of noise, stench, water, grey uniforms and mud" that formed his memories. He said he had "watched the white ships go sailing over, tiny specks in the blue, miles over his head, their serenity seemingly unruffled by the angry little puffs of smoke bursting around them." He observed, "The pilots were out of the sickening stench of the trenches, and had the chance of individual shows against the enemy. True, the danger was great—not only was there the Hun to contend with, but all the forces of Nature to reckon against—but the game seemed good."

Like Wilfred Blake, Ted Brown wanted out of the horror of war on the ground, preferring to take his chances under fire in a flimsy aircraft. In September 1915, he was transferred to the Royal Flying Corps as well. He was allocated to a squadron flying B.E.2c aeroplanes. These two-seat biplanes had been designed at the Royal Aircraft Factory at Farnborough, but they were manufactured by numerous companies across Britain. Some of them were built by Hilda Hewlett and Gustav Blondeau at their factory in Luton, north of London, which they set up after switching from tuition to aircraft manufacture as war approached. Their work quickly attracted a reputation for its high quality and for Hewlett's technical and organizational prowess. One article in 1915 noted, "Mrs Hewlett's machines were reported by the Admiralty to be far superior to those made by other firms." Another report observed, "Nothing but raw material is purchased. Every bolt, strainer, strut-socket, dome, engine plate, stamping and forging is made on the premises, and all tube-bending and woodwork is manufactured in the works." Hewlett's shops were spacious and well ordered, and her workforce was committed and hard-working. Nonetheless, it had been an anxious moment in

Hilda Hewlett (left) and colleagues working on a B.E.2c aeroplane for the Royal Flying Corps at the Hewlett & Blondeau factory, Luton, 1915.

early May 1915 when the young Fred Raynham took the first Hewlett & Blondeau B.E.2c's up for their test flights at Hendon, with Hilda Hewlett herself looking on.

As it turned out, there was no cause for unease. Raynham's tests were successful, and the aircraft were passed for the Royal Flying Corps. B.E.2c's were notably stable in flight. This made them good for observing. But they were less effective in the terrifying dogfights against agile German fighter planes that were increasingly becoming the lot of British crews over France and Belgium. B.E.2c's lacked speed and manoeuvrability. Royal Flying Corps crews made the best of what they had, though, and they were fearless; or, at least, the men kept their fear in check. After two weeks, Ted Brown qualified as an observer and began taking his seat at the front of the aeroplanes on their daily sorties from the squadron's aerodrome in northern France, near the border with Belgium.

He was just in time for the Battle of Loos.

———

THE BATTLE BEGAN ON SEPTEMBER 25, 1915, AND IT WAS A CHANCE for French and British forces to break through German positions fortified with barbed wire and machine guns. Royal Flying Corps crews flying overhead helped to direct Allied artillery fire—shells were in short supply—and to report on German troop movements. If the battle came off, it would be a major advance for the Allied forces.

The job of observer was an intense one and involved far more than its title implied. Wilfred Blake explained, "The observer is a trained wireless operator; he is an expert at photography under difficulties; he is a crack shot, and understands thoroughly the intricacies of machine-guns; he is a man of iron nerve and unblenching courage." The Canadian observer William Bishop wrote, "It is no child's play to circle above a German battery observing for half an hour or more, with your machine tossing about in air tortured by exploding shells and black shrapnel puffballs coming nearer and nearer to you like the ever-extending finger tips of some giant hand of death."

On Tuesday, October 5, Ted Brown and the Royal Flying Corps pilot William Allcock—he had no connection with Jack Alcock—climbed into their B.E.2c and took off. Brown had already teamed up with Allcock twice before. Together, they had carried out daring sorties near the enemy lines, with Brown observing up front, and Allcock skilfully piloting the craft from the rear seat. This time, they were to carry out artillery observation.

They were flying 8,000 feet above the little French town of Hulluch, twelve miles from Loos, when two anti-aircraft shells hit their fragile aircraft. Shrapnel from the shells pierced the engine, which immediately cut out. Instinctively, Allcock turned the damaged aeroplane towards the British lines.

Just as he crossed them, while they were still at 6,300 feet, the machine caught fire. Allcock pushed the nose of the aircraft downward in an attempt to land as quickly as possible, but in doing so, he caused the flames to spread along the fuselage. He and Brown were now sitting in a fireball as it hurtled towards the ground. Soon, the structure behind the pilot's seat was fully ablaze, and the front of the machine was so consumed by flame that Brown was forced to climb out of his seat and clamber backward towards his partner, gripping the thick wooden struts supporting the upper wing as he did so. Crouching on top of the flaming fuselage, he threw belts of machine-gun ammunition over the side to try to control the blaze.

Allcock continued to wrestle with the controls, pushing the nose farther down until the aircraft was diving almost vertically. By now, they were travelling at 120 miles per hour.

After what seemed like a lifetime, Allcock saw the ground approaching. From landmarks coming into view, he realized the aeroplane was headed directly for a small village near the town of Noeux-les-Mines. He pulled the aircraft, still in flames and with Brown clinging to the burning fuselage, out of its nosedive and around to the right. He spotted a field and aimed for its freshly ploughed earth. The machine sliced through a row of telephone wires strung along the edge of the field. Then Allcock levelled the craft out and hit the ground near the commune of Vaudricourt at 70 miles per hour.

It was a good emergency landing. But the fire had all but destroyed the aeroplane's undercarriage. After only a few yards, the B.E.2c collapsed, dug into the soil, and flipped over. Allcock was thrown out of the machine and landed, dazed, thirty feet away. The tail of the aeroplane, which broke off during the landing, smashed into his head, leaving him reeling.

Brown, who had landed underneath the aircraft's heavy engine, managed to roll clear only seconds before the engine snapped free of its mountings and fell, where it would surely have crushed him.

The pair reached each other and looked back at the aircraft. It continued to burn with such fierce heat that parts of the engine melted. Before long, the whole thing was destroyed. The men had survived with only burns and bruises, and they were behind British lines. Many had no such luck.

The airman Wilfred Blake, writing in 1918 of his wartime experiences, said, "Immunity from death may seem to be the luck of a pilot; he may shoot down many machines and be feared far and wide, but if he perseveres death overtakes him sooner or later; and all our pilots know this, and, knowing it, bravely face death a dozen times a day, not caring for themselves, but hoping that when it does come it will be quick—a bullet in the brain—and that they will be spared the sickening dive of the mortally stricken machine or the terrible torture of the flames."

On one occasion, a Royal Flying Corps aeroplane was watched by eyewitnesses as it dived, ablaze, to the ground. The observer jumped out, preferring the quick death of impact to the agony of incineration. His comrade remained in the pilot's seat, attempting in vain to bring the machine down safely. The ferocious flames, fanned by the dive, quickly consumed the floor of the fuselage, as well as the pilot himself. As the onlookers watched, his burning corpse dropped out of the aircraft to earth. Another time, two Royal Flying Corps aircraft accidentally collided at 3,000 feet above an aerodrome. Ablaze and locked together, the machines carried their burning pilots downward in a slow spin. An airman observing from the ground described the experience in his diary later that day. "God, it was a horrible sight," he wrote. "As I sat there watching, I kept trying to imagine what those poor devils were thinking about as they went spinning down into hell."

William Allcock and Ted Brown had escaped with their lives this time. But luck rarely lasts. Allcock was killed in a fierce air battle over northern France on June 5, 1917. It took over three months for official confirmation of his death to reach his grieving family. In a letter to his parents, the Royal Flying Corps chaplain described the young aviator as "a most skilful, as well as a dashing and dauntless, pilot." He was just nineteen years old.

A senior Royal Flying Corps officer, addressing a group of newly recruited airmen in October 1917, told them, "War is cruel, war is senseless and war is a plague, but we've got to win it and there's no better use of your life than to give it to help stop this eternal slaughter."

IN NOVEMBER 1917, AFTER MORE FIGHTING OVER THE FIELDS OF Europe and a lot more pain, Ted Brown returned to Britain and took a job at the Ministry of Munitions. He was to be a section director at the ministry's Department of Aeronautical Supplies, helping develop aircraft engines. There, he met Kathleen Kennedy. Kennedy had started her career as a General Post Office telephonist before becoming a commercial staffer at Lord Northcliffe's *Times* newspaper. Early in the war, she had joined the Women's Legion. Then she moved to the Ministry of Munitions, taking the job in the engines branch that introduced her to Ted Brown. She was twenty-one.

Both Kennedy and Brown had engineers as fathers. Arthur George Brown was the American Westinghouse man who got his son bitten by the engineering bug as a child. David Henry Kennedy had been born in Manchester but brought up in Gateshead, where Kathleen was born in 1896. He trained as an electrical and telephone engineer, reaching senior positions in Gateshead before moving the family to Ealing, in London, when Kathleen was a teenager. Like Ted Brown, David Kennedy enlisted in the Royal Flying Corps early in the war. In 1917, he had taken a technical job at the Department of Aeronautical Supplies. Like Brown's father, he had a string of engineering patents to his name.

With such similar parental backgrounds, Kathleen and Ted had much in common to discuss during their lunch and tea breaks. Soon, their friendship had grown into something deeper. Within weeks of Brown's arrival, the pair had become a couple.

While working in London, Brown also found time—perhaps encouraged by David Kennedy—to study air navigation. The topic had interested him as the war progressed, fomented, no doubt, by his aviation experiences in South Africa and then as a Royal Flying Corps observer. Besides grappling with complex theory, Brown went on to develop two navigational instruments, a position finder and a drift indicator, which David Kennedy applauded.

But the office life of the engines department didn't sit well with

Brown. His health was troubling him. He wanted to get back into the cockpit, but as a pilot, not an observer. In May 1918 he attended a medical examination in Hampstead, north London, where medics said that he needed open-air exertion rather than the indoor work of an office. They declared him fit to undertake flying instruction. The following month, he left the Department of Aeronautical Supplies. After a short break, he kissed Kathleen goodbye and joined the newly formed Royal Air Force.

Brown passed his flying tests on September 14, 1918. It should have been a moment of pride, and perhaps of relief after his confinement in the London office job. Anybody encountering him after flying had finished that day would have seen a man perfectly turned out, with an immaculate uniform, carefully brushed hair, and a neatly clipped moustache. But his face betrayed a troubled mind. He looked tense and distracted. His eyes stared into the distance as if he were recalling some memory better left forgotten. Maybe it was the moment he was shot down in 1915. Or perhaps it was something else, still more painful.

Ted Brown photographed after taking his flying tests, September 1918, aged thirty-two.

There seemed to be a darkness shrouding Ted Brown now. His wartime experiences—whatever they were—had knocked the youthful positivity from his spirit. Where once he saw a bright future, now he struggled to imagine if such a thing were possible at all. Perhaps there was no future. He clung to his relationship with Kathleen as a crutch. Her love restored a measure of meaning to his existence.

Brown took up a posting as a flying instructor at a Royal Air Force base in Lincolnshire. In his new role he flew numerous types of aircraft, spending most of his time in those made by the firm of Handley Page. Separated from Kathleen except for short periods of leave, and missing her like mad, he asked for her hand in marriage that October, and she accepted.

FRED RAYNHAM AND HARRY HAWKER WERE PRESSED INTO WARtime service as test pilots. Raynham, who was said to have "an insatiable appetite for such work," continued to fly for Avro. He also freelanced for other manufacturers, and tested hundreds of scouting biplanes made by Martinsyde, which were in high demand at the front. Hawker continued to test the Sopwith firm's prodigious output, as well as contributing design work as new aircraft were commissioned.

Hawker always wanted to push the boundaries of what was possible in the air. In June 1915 he broke the British altitude record, climbing to over 18,000 feet. The following year he broke all world records by reaching 23,600 feet—four and a half miles above the Earth's surface—in a Sopwith biplane over Brooklands. He took no oxygen, nor was his aircraft specially modified, and it was a hard flight. "The cold, of course, was intense," reported *Aeronautics* magazine. It said that Hawker experienced "a sensation of extreme lassitude which rendered the slightest exertion or movement almost impossible." Frostbite attacked his fingers. But it was his dramatic return to earth that most astonished those looking on. He did it by "tipping his machine into a nose-dive and coming down like a rocket-stick," a later newspaper report noted. It concluded, "Only a man in the prime of physical fitness could have stood such a lung test."

Some of Hawker's work was carried out away from public view. The Sopwith company worked closely with the Royal Navy on plans for launching and landing aeroplanes on the decks of moving ships. *Campania*, an old Cunard passenger liner rescued from the shipbreaker's yard, was modified for trials. Long in the tooth the liner might have been, but it was big and fast. The first successful flight off its moving deck took place in August 1915, when a small Sopwith Schneider biplane, lightweight but with a powerful engine, was piloted off after a 113-foot run. Landing on a moving ship was a harder task. The air forced up in the wake of a warship moving at speed—"the bump that the albatross sits on," as one expert described it—was challenging and unpredictable to fly through. It was only two years later, in August 1917, that a little Sopwith Pup was successfully brought in to land on the moving deck of *Furious*, a newly built aircraft carrier.

By then, Britain's navy needed to deploy heavier aircraft, capable of flying long reconnaissance missions as well as carrying enough ammunition to attack German Zeppelin airships high overhead. In November 1917, just such an aircraft—a Sopwith 1½ Strutter—was flown from the moving deck of *Campania*. By then, the vessel was under the command of a thirty-seven-year-old naval navigator, Kenneth Mackenzie Grieve, whom Hawker knew as Mac. It was a challenging exercise. The aeroplane had been heavily laden. Besides its crew of two, it carried enough fuel to fly for eight hours. It needed a long and careful take-off, challenging in the best conditions. For the deck crew, preparing the aircraft was no easy task. The Sopwith's removable wings, stored on top of the aircraft below the deck, had to be attached to the fuselage in minutes, despite the wind blowing across the deck, which tried to wrest the wings from the men's grasp. But Tom Sopwith and Harry Hawker had done their work well, and *Campania*'s crew, under Mac Grieve's careful command, proved more than equal to the task. Harry and Mac would work together again, though they didn't know it at the time.

ON NOVEMBER 14, 1917, FOUR DAYS AFTER THE SUCCESSFUL LAUNCH of the heavy Sopwith from *Campania*'s deck, Harry Hawker married. His bride, twenty-two-year-old Muriel Peaty, had grown up in a household which, like Hawker's, prized hands-on activity. Her father was the UK managing director of Pleyel, Wolff, Lyon, a leading French piano and harp manufacturer, and ran its busy London showroom on Baker Street. He also held a fellowship of the Royal Philharmonic Society and was well known in the music performance world. Muriel's brother, Leonard, two years her elder, was crazy about motoring. He studied engineering at the University of London, then got a job in the automobile workshops of the Clément Motor Company. When the war broke out, he joined the Royal Flying Corps to work on aeroplanes instead, but not before he had seen his little sister get caught by the motor car bug as well.

Muriel and Harry met during the early years of the war. Hawker had developed a habit of going for Sunday drives with a friend through Richmond Park, close to the Kingston factory where Sopwith built its aircraft. So, too, would Peaty. One Sunday in April 1915, Peaty and her companion had broken down halfway between the park's Kingston and Richmond gates. After examining her car's engine, Peaty diagnosed the problem, but to fix it, she needed a top-up of petrol. Hawker had been passing by. Stopping, he offered some spare fuel he kept in a tin in the back of his big Grégoire motor car.

Once Peaty's car was fixed up, she and Hawker swapped telephone numbers, though Peaty didn't expect anything to come of it. Hawker had seemed cool towards her throughout the encounter and appeared glad when she had finally driven off. She was surprised, therefore, to receive an enthusiastic telephone message from him the following Sunday morning. He wanted to invite her and her friend to help him test a new car he had just bought. It was an Austro-Daimler: fast, powerful, and a lot of fun to drive. He appreciated Peaty's own interest in cars, and wanted her to try it as well. She always considered it the first real motor car she had driven.

These Sunday drives took place each week for the rest of the summer. They usually continued to the nearby Brooklands track, where Hawker often tested aircraft. He souped the Austro-Daimler up until it could race along at 80 miles per hour, four times the speed of most vehicles on the road in those days. Soon enough, he met Peaty's parents, and they got on well. Peaty secured a job at a government agency with offices near Buckingham Palace. The day she was offered the job, she and Hawker became engaged.

Both were strong-willed, and though their bond was tight, their early relationship was, she later admitted, "stormy." Hawker's friends could hardly believe he was engaged. One of them remembered, "We just could not imagine him taking time off to get acquainted with a girl."

By the time the pair married in 1917, a further dark cloud was looming on the horizon: a secret about Hawker that they both came to share. Just days before the wedding, Hawker was delivering an aeroplane to Villacoublay, in France, when bad weather forced him to land on the way. The aircraft got stuck in thick snow. Hawker spoke no French and was dressed in civilian clothes. The French authorities, suspicious, arrested him. He needed to get the aircraft to its destination before darkness fell. With time running out, he managed to talk his way out of custody, then waved down a passing English lorry. With the help of its two occupants, he freed the aeroplane from the snowdrift, taxied, and took off. He made it to Villacoublay in time, but it came at a cost. As he was pushing the machine from the snow, he twisted his back, worsening an injury he had suffered four years earlier in an accident that almost cost him his life.

Hawker had been attempting to win another record-breaking prize on October 8, 1913, when he crossed the Brooklands aerodrome too low in strong winds. He was caught in a gust, with no altitude to recover, and dived to the ground. One of his aeroplane's wings hit the bank of the River Wey, just outside Brooklands, and the machine crashed. The impact wrenched Hawker's back and landed him in hospital. Three weeks later, seemingly recovered, he again tried for the prize. This time, after three hours of flying, he was forced to abandon the mission with a blinding headache. In June 1914, during an air race from Hendon to Manchester, he took seriously ill while at the controls. He managed to land at Hendon safely, but he was almost unconscious. A week later, after successfully performing a loop-the-loop at Brooklands, he fell into a spiral nosedive. The machine crashed on its right wing in a coppice of trees. It was an "extraordinary escape from death," said one eyewitness.

Troubled by his bouts of pain and sickness, Hawker nevertheless refused to be held back as he spent the war working for Sopwith. But the back injury in France in 1917 had been the worst he had yet experienced. After a month of agony, he sought advice from his doctor, who told him to rest in bed until the injury healed. Two more weeks of pain and frustration yielded little improvement, so Hawker decided to get up and work through the pain instead. For a while, it seemed to do the trick. But it wasn't a cure.

As the war progressed, the sharpness inherent in Hawker's temperament had become more evident to those around him. His easy charm had more often given way to impatience or unreasonableness. A young Australian aviator, Horrie Miller, knew Hawker well during the war. He later described Hawker's reaction to the crash of 1913 that put him in hospital. "He made light of this, and some time later when he was warned of a possible spine infection that could lead to paralysis, he told no one."

CHAPTER 5

CRASHED

BY THE TIME WAR WITH GERMANY WAS DECLARED, TWENTY-ONE-year-old Jack Alcock had become one of Britain's most experienced and accomplished flyers. He joined the Royal Naval Air Service, the aviation wing founded to support naval operations, in November 1914, and for the next three months, he travelled across Britain and into mainland Europe to examine aeroplanes and instruct pilots. In February 1915, he settled at the Royal Naval Flying School at Eastchurch, on the north-east Kent coast, as senior flying instructor.

"Men and machines were badly needed then," he recalled, "and with others equally keen I set to work in order to produce the necessary pilots for the Air Service. It was hard work, and difficult at times owing to the serious shortage of machines, but gradually order was derived from chaos, and we turned out some of the cleverest pilots on any front."

Alcock's experience at Eastchurch was shaped in large part by its location. It was an isolated, desolate posting for a young man more used to the bustle of city life in Manchester or the activity on busy aerodromes around London. Eastchurch, an airfield since 1909, sat in the middle of Kent's cold, marshy Isle of Sheppey. Its proximity to the North Sea coast made it a regular target for German airship bombing raids as the Zeppelin fleet geared up its campaign over England that spring. The Zeppelin airships, sometimes called "sausages" because of their elongated shape, were becoming a menacing presence over Britain's shores. Only days into the job at Eastchurch, Alcock told a Manchester friend, "I shall be very pleased when the war is over, as it is a very trying ordeal to be under shrapnel and shellfire in an aeroplane." He said, "The last two nights I have had to spend in an arm chair, owing to the German sausage's activity round our coast." By the summer, German air raids had pockmarked the Eastchurch aerodrome with craters.

As 1915 wore on, Alcock settled into his work. He was up in the air with pupils from sunrise to sunset every day. At a performance review a year into the job, his squadron commander judged him "one of the steadiest pilots in this country." His hard-working ethic was soon rewarded with promotion to flight sub-lieutenant. Countless men passed through his careful hands, including Charles Morgan, a twenty-five-year-old former naval midshipman, and Herbert Brackley, who arrived at Eastchurch aged twenty from a journalism job at the Reuters news agency. Like Harry Hawker with Mac Grieve, Alcock was to meet both Morgan and Brackley later, though none of them, in the depths of war, could have imagined how or where.

By the end of 1916, after nearly two years of instructing at Eastchurch, Alcock had racked up two thousand hours of flying time on all kinds of aircraft. He was earning the highest plaudits from his superiors. They described him as an "excellent instructor" and a "brilliant pilot," with "quite exceptional ability." But he was getting frustrated. He had become convinced that the best way he could serve his country at such a defining moment in its history was not to teach, but to fight. He wanted to leave the confines of Eastchurch and seek a posting to one of the overseas fronts.

Alcock later observed of his efforts at Eastchurch, "It was useful work, and very necessary also, but we chafed at its limitations. The desire to be out and in it was always with us. To get a taste of the real thing obsessed every one." Another Royal Naval Air Service pilot, just a year younger than Alcock, had also spent the first part of the war in Britain. In a letter home in early 1915, he wrote, "I wish we could get out to the front." He admitted, "I would much rather come to a sticky end out there than here." It was the same for Alcock.

By 1916 the Eastchurch base had grown sizeably, accommodating upwards of a thousand people. But for ambitious flyers like Alcock, it always felt remote. Even its commanding officer was forced to observe, "Eastchurch was never a very lively spot and it was even less so in war time, particularly in the middle of winter." In December 1916, Alcock's request for a posting overseas was granted. Along with it came a much-prized promotion to flight lieutenant.

ALCOCK EXPECTED TO TRAVEL TO ROMANIA, BUT AT THE LAST minute he was directed south to the Allied naval base at Mudros. Mudros sat on the Greek island of Lemnos, in the Aegean Sea near the mouth of the Dardanelles Strait. The facility had been set up as the storage and maintenance base for aeroplanes stationed across the Aegean. It also housed airships and seaplanes carrying out anti-submarine patrols. *Ark Royal*, Britain's first purpose-built aircraft carrier, by then acting as a support ship for local seaplane operations, was also based there.

Alcock's arrival at Mudros, as the new year of 1917 dawned, immediately intensified his experience of the war. He recalled, "One got all that one could wish of fighting." The aircraft he encountered at Mudros were largely familiar. They were mostly Sopwith scouting aeroplanes that he knew well from his Brooklands days with Harry Hawker. Alcock used these small aeroplanes daily to carry out repair, supply, and anti-submarine work with his Mudros colleagues and with those on neighbouring Aegean island bases such as Imbros. They also carried out bombing raids on Turkish targets, though the small size and carrying capacity of the aircraft limited their effectiveness.

But the scope of their operations was soon to change. In May 1917 Mudros took on a new role. With the United States having entered the war, and Russia convulsing in revolution, Allied commanders planning the next phase of fighting in the Dardanelles decided to take a more aggressive line. They wanted to start attacking targets deep within Turkish territory. Railway lines, naval destroyers, and enemy command centres as far away as the Turkish capital of Constantinople were to be targeted in this new offensive. But long-range attacks could only be carried out with a new type of aircraft. To perform the raids, Mudros took delivery of a brand-new twin-engine Handley Page heavy bomber, capable of carrying big payloads of high-explosive bombs over long distances. Its arrival was to change the direction of Jack Alcock's life.

THE FIRST BOMBING ATTEMPT ON CONSTANTINOPLE WAS MADE ON July 3, when a detachment of nine Sopwith bomber and fighter aircraft approached the Gallipoli Peninsula. Their target was the shipping moored in Gallipoli's inner harbour, as well as warehouses by the water. The daylight raid was a success: bombs managed to hit four warehouse targets. But it was also a diversion. As the Turkish authorities were dealing with the aftermath of the raid, Mudros's new Handley Page bomber was being readied for its first night attack. After darkness had fallen, it was flown over Gallipoli, up the west coast of the Sea of Marmara, towards Constantinople. But engine trouble caused by the hot weather forced the crew to turn back. With a heavy payload of bombs and the engines rapidly losing power, the aircraft almost crashed into the water. To lighten the load, its lead pilot, twenty-two-year-old Kenneth Savory, released six bombs, two of which hit an enemy camp about one-third of the way down the peninsula. It was enough to let the aircraft level out, and it limped back to base.

Five days later, on July 8, the bomber made another night attempt on Constantinople. While the smaller Sopwiths made fresh raids along Gallipoli to draw attention, the Handley Page crew flew over the Sea of Marmara a second time towards their target. This time, a strong headwind forced them to turn back. It was another frustrating setback.

The next night, July 9, a third attempt was made on Constantinople. This time, Savory and his crew made it. Just before midnight, after three hours of flying, the heavy bomber arrived over the city's congested harbour area. Flying at just 800 feet, Savory spent over thirty-five minutes reconnoitring the area. Then he dropped a series of eight 112-pound bombs on the shipping and facilities below, hitting a destroyer and leaving it ablaze after a large explosion. Two further bombs hit the German military headquarters. Finally, the crew released its remaining pair of bombs on the Turkish War Office, each one scoring a direct hit.

This daring midnight raid on the heart of Turkey's capital, which struck military targets across the harbour area, left the enemy authorities stunned. It had come as a complete surprise. By the time the Turkish military activated its local searchlights and air defence guns, the raid was all but over. As Kenneth Savory turned the Handley Page for home and headed back along the Sea of Marmara, the Turks fired a furious barrage of shells in its direction. One piece of shrapnel damaged an engine, and the rest of the aircraft took a battering. But they made it away and landed safely at Mudros at 3.30 a.m.

Now that the Turks had realized that Mudros housed a long-distance bomber capable of reaching Constantinople, their own attacks on Allied forces were stepped up. On July 27, eighteen days after the successful attack on Constantinople, the island of Lemnos was surveyed by three German seaplanes carrying out reconnaissance in preparation for retaliatory strikes. As the hostile aircraft made their way to Mudros at a height of 7,000 feet, five of the base's airmen scrambled fighters to intercept them. Only Jack Alcock, in a Sopwith Camel, had the power to reach the incoming German formation.

The German pilots never saw Alcock coming. As they flew over a small island near the mouth of the Dardanelles, Alcock darted towards one of the escort fighters. He had manoeuvred himself so that he had the sun behind him. He approached the fighter with his finger locked onto the trigger of his twin machine guns. The German pilot was blinded by the sun. He could do little to avoid Alcock's point-blank gunfire. Then, to avoid colliding with the German machine, Alcock looped underneath and approached it from the tail, firing another long burst from his guns. It was a direct hit. The fighter dropped out of the sky towards the sea just off Sedd el Bahr, on the nose of the Gallipoli Peninsula.

As the damaged German seaplane dived downward, its companion escort followed it. The pair levelled out close to the water. Alcock stayed hard on their tail. He chased after them, firing heavy bursts of gunfire.

Suddenly, one of his guns jammed. He kept a cool head. While keeping the Camel on course to its target, he cleared the jam, then renewed his attack on the stricken seaplane. This time, he finished the job. The aircraft nosedived into the water.

Alcock then swung the Camel towards the second fighter, giving it little chance to escape. Getting close, he fired. It was another hit. The last Alcock saw was the damaged seaplane disappearing low over the mainland of the peninsula.

The third machine in the enemy formation, the reconnaissance seaplane that the fighters were protecting, escaped Alcock's guns. While Alcock was attacking its escorts, it managed to land on waters protected by Turkey's coastal artillery batteries before Alcock could reach it. Nonetheless, it was a humiliating result for the Germans and Turks, still smarting from the Constantinople bombing. It also gave Alcock valuable experience he would call on two months later, when the island was again reconnoitred for an attack.

IT WAS SEPTEMBER 30. LOOKOUTS ON *ARK ROYAL* REPORTED THREE German seaplanes approaching Mudros. A two-seat Friedrichshafen reconnaissance aircraft was escorted by two single-seat fighters that the British called Blue Birds—they were probably Albatrosses or Rumplers. As word was being passed to the airmen on land, the ship's gunners swung their twelve-pound anti-aircraft gun towards the German seaplanes and let off three rounds. The explosions filled the sky with bursts of heavy shrapnel. One of the Blue Birds was struck, causing damage to a supporting strut. Nonetheless, it continued to fly.

When the order to intercept the incoming seaplanes reached the Mudros airbase, Jack Alcock was taking his morning bath. He jumped out and dried himself off. Instead of finding his clothes, he got into his pyjamas, then ran outside to a Sopwith Camel waiting on standby. As he did so, a fellow pilot, eighteen-year-old Harold Mellings, raced towards a Sopwith Triplane. The pair climbed into their respective cockpits. Both aircraft were started by the ground crew. Within moments, the two airmen taxied their machines down the airfield and took off into the lightening morning sky.

The German seaplanes were flying at a height of about 12,000 feet over the Lemnos coast, with the Friedrichshafen in front and the smaller Blue Bird fighters flanking it in the rear. It must have seemed like a re-run of the July attack that Alcock had successfully fought off. This time, though, he was not intercepting the hostile aircraft by himself. As he pushed the Camel in a steep climb towards the enemy machines, he saw Mellings edge ahead in the Triplane.

Mellings was first to attack. He approached the Friedrichshafen head-on, then fired the Triplane's powerful machine gun. As he did so, one of the Blue Birds dived downward, under Mellings's craft, before turning and firing at him from behind with its twin guns. Mellings, experienced by now in the art of evasive manoeuvres, darted out of reach. Heading now towards the second Blue Bird, he opened fire. It was a miss.

Just then, Alcock fired on one of the Blue Birds with both of the machine guns mounted on his Camel. He had got close to his prey—just fifty yards. With such a bombardment of gunfire, he scored a hit. Smoke poured from the machine. Alcock couldn't tell whether it was engine damage or a smokescreen. Then the Blue Bird dived away.

Mellings had now manoeuvred the Triplane behind the other Blue Bird. At 100 yards, he opened fire, and kept his finger on the machine gun's trigger as he closed in on the enemy machine. The Blue Bird quickly dropped away, attempting to escape Mellings's hail of bullets. Mellings followed, chasing his target down towards the water.

As the pair approached the sea, the Blue Bird pilot pulled out of his dive and flattened out, skimming the waves of the ocean as he tried to evade Mellings's attack. Mellings stuck to the German's tail. Racing along just twenty feet above the water, he let loose a further burst of rapid gunfire. His bullets ripped through the Blue Bird's upper-left wing, tearing off a large section. One bullet then struck the German pilot in his back. By now, Alcock had followed the pair down to the water and was ready to support Mellings if he needed it. But the job was finished. With a strong wind speeding the stricken Blue Bird along at 100 knots, it slammed into the sea, breaking up into small fragments which sank immediately. The pilot was lost.

With no time to celebrate his hit, Mellings pushed the Triplane back up to the dogfight and made for the two-seat Friedrichshafen. Meanwhile, Alcock returned to the remaining Blue Bird, firing all the while. At their closest, the two fighters were just six feet apart. Alcock's bullets riddled the German's petrol tank and buoyant floats. As he pressed forward, chasing the aircraft down, Alcock succeeded in pushing it towards the neighbouring island of Imbros.

Then, without warning, both of Alcock's guns jammed. Mellings, in turn, had been firing on the Friedrichshafen. His gun jammed at exactly the same time. Now he was to feel the heat of attack as the seaplane's observer unleashed a hail of bullets towards him.

Both Mellings and Alcock were now unarmed and easy targets for the two remaining German machines. Alcock, seeing that Mellings was under attack, left the Blue Bird he was tailing and launched his machine towards the Friedrichshafen in an aggressive manoeuvre as if he was about to open fire. The feint worked. The larger seaplane, separated from its escorting fighters, feared an imminent attack. As its pilot carried out evasive manoeuvres, his observer momentarily stopped firing on Mellings. This gave Mellings just enough time to clear his jammed gun. Thanks to Alcock's covering attack, he could now once more shoot at his adversary. Alcock returned to the Blue Bird.

At that moment, Mellings was joined in the sky by Philip Fowler, a twenty-two-year-old New Zealander, who joined the fray in a Sopwith Pup. While Mellings shot at the Friedrichshafen's nose from just twenty yards, Fowler fired from the rear. He hit and wounded the observer.

It looked as if the Friedrichshafen was done for. The observer was hurt, and the pilot was struggling to shake off his attackers. Then the fight turned in the Germans' favour. By this point, the Triplane, the Pup, and the Friedrichshafen had reached the Dardanelles. Now the German pilot brought his craft down low over the water, putting it within reach of the Turkish artillery forts protecting the region. If Mellings and Fowler wanted to continue the assault, they would need to descend as well, and risk being caught in shellfire. It was a chance they were not prepared to take. They broke off their attack and made for home. The reconnaissance seaplane and its wounded crew landed in safety.

Meanwhile, Alcock continued to chase the remaining Blue Bird, even though his guns were still jammed. In trying to get away from Alcock's approach, the German pilot, later identified as J. R. Muller, banked vertically downward. As he did so, the supporting strut that had been damaged by anti-aircraft fire from *Ark Royal* snapped under the strain. Both of the Blue Bird's left-hand wings folded up as the machine fell. Unable to fly much farther, Muller was forced to attempt a landing on the choppy waters of the Aegean. But Alcock stayed hard on his tail. The impact as the Blue Bird hit the water was so violent that both its floats were torn off, crippling the machine. Almost immediately, it began to sink, and quickly slipped beneath the surface. All that remained was one of the floats.

Pilot Muller, who was wearing an inflatable life jacket, was left drifting in the cold water of the Aegean, clinging in desperation to the float. Fearing the worst, he fired a distress flare into the sky. Alcock was circling the Camel around the site of the impact, preparing to go to the German airman's rescue, when his engine started misfiring. He quickly climbed and began making for the nearby Imbros base, hoping to reach safety before the engine cut out. He made it.

After Alcock reached Imbros and filed his report, he took another aircraft up and began searching for the German pilot. There was no trace of him to be found. Mid-morning, a Mudros-based British destroyer, *Acheron*, was diverted from its daily patrol to look for him. For over three hours, *Acheron*'s crew searched the waters of the Aegean. As they did so, Muller, unaware that rescue missions had been launched, had grown increasingly desperate. After several hours drifting in the water, he scratched a short message into the thick green paint of the Blue Bird's float. Addressing his girlfriend, he wrote, "I am dead, tell my parents." He pulled out a pocketknife and punctured the inflatable jacket that was supporting him in the water. Then he began to cut into his right wrist.

It must have been at that moment that Muller spotted *Acheron* in the far distance. He abandoned his attempt at suicide. Then he waited, clinging to the float, as the British ship approached. At 1.30 p.m. he was pulled out of the water by *Acheron*'s crew and taken on board. His wound was patched up and he was transferred to a smaller ship while *Acheron* continued its patrols. Muller was now a prisoner of war, but he had survived. His interrogation yielded useful intelligence for the British. Much later, his girlfriend would write to Alcock, expressing her respect for his fighting skill. She observed that Muller was "one of the best," and said that Alcock "must have excelled him to have beaten him down."

WHATEVER THE OUTCOME OF A DOGFIGHT, SHOOTING DOWN AN aeroplane could be an affecting experience. One fighter pilot, William Bishop, recalled the moment in March 1917 when he shot down his first German aircraft. Writing only the following year, he said, "It made my heart leap to see my smoking bullets hitting the machine just where the closely hooded pilot was sitting." Then Bishop explained how he had followed the aircraft as it dropped towards the ground with its pilot either mortally wounded or dead. "I went into a wild dive after my particular opponent with my engine full on," he recalled. "When I was still about 1,500 feet up, he crashed into the ground below me. For a long time I had heard pilots speaking of 'crashing' enemy machines, but I never fully appreciated the full significance of 'crashed' until now. There is no other word for it. I have not to this day fully analysed my feelings in those moments of my first victory. I don't think I fully realized what it all meant." Another time, Bishop managed to creep up on an enemy aircraft unnoticed. He fired twenty bullets into its pilot, killing him instantly. "That pilot never knew what happened to him," Bishop recalled. "Death came to him from nowhere."

If Jack Alcock had been affected by his experiences in dogfights, he kept his feelings to himself. His action in July earned him high praise from his commanding officers. The September battle secured him a Distinguished Service Cross, which he later described as "much prized." His wing captain said that Alcock had shown a "most praiseworthy balance of judgment and unselfishness."

Alcock was now an experienced and versatile pilot. Bombing raids carried out in the Handley Page had increased in the weeks that followed the July dogfight off Sedd el Bahr. When pilot Kenneth Savory was redeployed, Alcock was given command of the heavy machine. On August 4, a week after he had downed the two German fighters single-handed, he piloted the big bomber on a night raid to the Turkish port city of Panderma, dropping bombs on warehouses and storage dumps along the coast to devastating effect. A huge explosion triggered by one of the bombs caused the Handley Page to rock violently as it flew away. On September 2, with Alcock again at the controls, the Handley Page bombed the railway station, a railway bridge, and a Turkish fort at Adrianople. Warehouses in the industrial district near the railway station were also hit, causing multiple explosions and widespread fires. As he flew back to base, Alcock could still see the glow of the flames lighting up the night sky when he was ninety miles away from the city.

Alcock's work at the controls of the Handley Page bomber proved his versatility. These long-distance flights, lasting several hours at a time, gave him a taste for a type of flying that was quite new to him. Once the war was over, he would want to put that experience to use.

AFTER FOUR WEARYING YEARS, THE END OF THE WAR FINALLY came. Addressing Britain's Parliament, the king gave thanks to those he described as "the keen-eyed and swift-winged knights of the air, who have given to the world a new type of daring and resourceful heroism." The Armistice, when it was signed on November 11, 1918, prompted a riotous celebration.

That Christmas was like no other. Those who had survived the war made their way home—whatever home they had. Jack Alcock, by now aged twenty-six, arrived at Dover, on England's south coast, on December 16. He had endured a long and gruelling journey back from the Aegean. It was just in time, he said, for "a Christmas which some of us had never expected to see again in our native land."

Fred Raynham, aged twenty-five, must have been relieved to see the end of wartime flying. He had suffered a serious injury halfway through the war. While testing an aeroplane for the Royal Flying Corps at Brooklands, its tail broke off in mid-air. The aircraft fell from a height of 2,000 feet. Raynham was found unconscious, his face a mask of blood. He had been taken to hospital in critical condition. If the aircraft had hit the concrete track or the extensive buildings occupying the airfield, death would have been certain.

Harry Hawker, at twenty-nine the eldest of the Brooklands trio, had settled into married life with Muriel. In October 1918, their first child, a girl, had been born. They named her Pamela. For the Hawkers, the Christmas of 1918 was a time for family and for making optimistic plans.

Mac Grieve, the thirty-eight-year-old commander and navigating officer of the aircraft carrier *Campania*, approached Christmas 1918 with trepidation. That November, *Campania* had been lying at anchor in the Firth of Forth. On November 5, six days before the Armistice was signed, a storm caused the ship to break free. After it collided with other Royal Navy vessels nearby, its hull was breached, and it began to take on water. A few hours later, it sank. Its masts and funnels remained visible above the surface of the firth. Its precious cargo of eleven naval aircraft, seven of them Sopwiths, was forever submerged. It was an ignominious end for the pioneering vessel, and for Grieve's wartime service.

For Ted Brown, now aged thirty-two, it was a particularly low time. The ill health he experienced working at the Department of Aeronautical Supplies in London had not improved by his relocation to Lincolnshire, where he gave flying instruction in the dying days of the war. By December he was, in his own words, "not at all fit." A medical exam that month recommended his discharge from the armed forces and ordered him home to his parents in Manchester. Before he could leave the aerodrome, his fragile state of mind suffered a further shock. A little terrier belonging to another officer ran up to an aeroplane that was readying to take off. Excitable in the frosty winter air that day, the dog jumped up and bit at the aircraft's whirling propeller. The sight was sickening. Brown had to shoot the dog dead.

Christmas was spent on enforced sick leave. With Kathleen still working in London, Ted could only communicate with her by letter. The pair were devoted correspondents. Ted wrote to Kathleen often, sometimes daily, and she replied to each letter straight away. He addressed her with nicknames that were affectionate to the point of soppiness. To her, he was always "Teddy." His letters were demonstrative and tender, as if they had been written by a boy in love for the first time. Perhaps he was. Kathleen could have been in no doubt as to how much he adored her—and needed her. He showed a sentimental and vulnerable side of his character that outsiders never glimpsed. He sometimes admitted to Kathleen that he was lonely and depressed. He told her that he could see no beauty in the world unless she was with him to share it. One time, he begged her to help him hope for the future.

It was a frustrating and unsettling time for the couple. Writing back and forth into the new year of 1919, Ted and Kathleen wondered to each other what the next chapter of their lives would hold.

AS IT TURNED OUT, THE ANSWER FOR TED BROWN WAS THE SAME as it was for Jack Alcock, Harry Hawker, and Fred Raynham, though none of these very different characters knew it at the time. The *Daily Mail* transatlantic flight competition, first announced by Lord Northcliffe in 1913 but postponed by the war, was back on. What a difference the intervening years had made. The first aviators had been born into a time of great hope. Too soon, those pioneers had been forced to grow up, rapidly and painfully, in the most brutal of wars. For those who survived, now was the moment when they might come of age. In that short time, aircraft themselves had developed beyond all recognition as well. Fragile experimental craft made by amateurs in garages and sheds from the slimmest struts and the cheapest linen cloth, held together as much by hope as by science and seeming to fly only according to whim, had been replaced by sturdy and reliable machines designed and built by experts, hardened by the demands of aerial warfare and powered by good, strong engines.

In the weeks and months following the Armistice, aircraft constructors, mechanics, engineers, pilots, and navigators would coalesce into teams. Each would build a machine they hoped would be the first to make the Big Hop across the Atlantic Ocean. There was much optimism. There was also a sense of urgency: once the hostilities had ended, aircraft manufacturers needed new customers. It would be a huge publicity coup for the first to cross the Atlantic nonstop. But everyone knew it would be the most difficult journey they could ever make.

PART TWO

Aviators do not know what fear is. When you are up in the air the exhilaration of flight and the necessity for careful control of your machine takes away all other feeling.

—HILDA HEWLETT, AVIATOR, 1912

CHAPTER 6

SLEEP AWAY THE NEXT TWO MONTHS

MURIEL HAWKER NEVER FORGOT THE MOMENT HER HUSBAND brought news that changed the course of their lives. "One evening, soon after the Armistice," she recalled, "Harry came in and said he had been asked to fly the Atlantic with a machine which Sopwith's were prepared to build. He had always been keen on the flight, and I knew it would come sooner or later."

After their wedding, Muriel and Harry Hawker moved into a villa in the middle of Hook, a postcard-pretty village in Surrey. Their new home sat on a generous one-acre plot opposite the village church, a fifteen-minute drive from the Sopwith factory in Kingston and only twenty-five minutes from Brooklands. Its garage became home to the couple's growing stable of motor cars, which comprised the Grégoire that Harry had been driving when they first met, a Talbot saloon, and a sporty Ford Model T. There was also room for a large workshop adjoining the garage. Here, the couple embarked on a project to build their own sports car to add to the collection. It would be constructed on a chassis from a Mercedes car, designed for use with a thirty-five-horsepower engine. But the Hawkers had something a bit bigger than that in mind. The car they planned to build would be powered by a twelve-cylinder, 225-horsepower aircraft engine sourced from Jack Alcock's former employer, Sunbeam.

It was an absorbing project for the pair as they settled into the first few months of married life. After modifying the chassis and its gear-box to accommodate the huge engine, Harry then shaped aluminium sheeting to form a seven-foot-long bonnet. Muriel riveted the completed sheets together. The rest of the bodywork was obtained from a specialist supplier to Harry's design. When they first started the unsilenced engine inside the garage, the roar made the walls shake.

After it had been completed to the couple's satisfaction, the Sunbeam car proved to be a beast of a machine. They went for long, fast rides in it whenever Harry had a break from testing aircraft, giving them pleasure and release as the final months of the war passed by. But the birth of their daughter in October, less than four weeks before the war ended, brought a new responsibility into their lives.

When Harry told Muriel about the transatlantic flight opportunity offered by Sopwith, her first thought was for baby Pamela. The two-month-old was about to experience her first Christmas. It was hard to countenance any activity that might risk her father's life, as well as the family's livelihood. But Muriel knew how much the flying competition meant to her husband. She recalled, "I was torn between my duty to Pam to ask him to stay and my duty to him to let him go. I tried to imagine how I should feel if another man were to fly the machine that Harry ought to fly, just because I feared the consequences. I knew I could never allow that to happen."

Harry understood the predicament in which he had placed his wife. Theirs was a partnership of equals, and this was not a decision he would make alone. He told Muriel that if she didn't want him to take part in the competition, he would withdraw—and he would understand. But Muriel had already made up her mind. She told him, "Why should you think I want you to stay? I want to be proud of you."

Too soon, Christmas was over, and the new year of 1919 arrived. In Paris, world leaders began to thrash out the terms of peace. In Ireland, the war for independence began. A global influenza pandemic, which had started the previous year, continued to ravage. The relief and elation experienced with the ending of the Great War was soon tempered. The

Daily Mail transatlantic prize offered a much-needed distraction to those who decided to compete, and to the wider public looking on.

For Harry Hawker, as for everyone working in Britain's aeroplane industry, peace paradoxically brought a sense of unease. Since 1914, every aircraft the factories could turn out had found a ready buyer. The war's appetite had been insatiable. It was a different matter in peacetime. The signing of the Armistice saw Sopwith's business collapse. Tom Sopwith later said, "It wasn't a question of winding down. It was turning off the tap. We had scores of sub-contractors. I really can't remember how many—and, suddenly, overnight, no one wanted any more aeroplanes." Existing orders were cancelled. The company's workforce was cut from 5,000 to 1,400. The workshops were emptied and surplus military aircraft, many of them of little use in the civilian world, were sold off at rock-bottom prices. A market for civilian air travel was yet to be built. Hawker knew how much was resting on the *Daily Mail* contest. Sopwith's future as an aircraft manufacturer, and his own career as an airman, might depend on the publicity for aviation that crossing the Atlantic would bring their way. He was determined to be the first to try. He told his sister, "I am getting my kick in before the others start."

In January, Hawker and the Sopwith team started work. They gave themselves just two months to design, build, and test a brand-new aeroplane capable of flying the Atlantic in one jump. They would be carrying out the work in secret. And they had a new member of the team to bring up to speed, though he was already familiar with Sopwith aircraft, and with the firm's staff. Mac Grieve, the naval commander who worked with Sopwith on *Campania* during the war, had been chosen to join Hawker as navigator on the transatlantic flight.

KENNETH MACKENZIE GRIEVE, KNOWN AS MAC, WAS BORN INTO A naval family in 1880. His father, who grew up in Edinburgh, had joined the Royal Navy at the age of fourteen and steadily climbed its ranks, being decorated for action in the Anglo-Egyptian War and the Nile

Expedition for the relief of General Gordon at Khartoum. He and his wife raised Mac and their other children in a determinedly Christian household with a strong sense of duty. Mac followed his father's example by joining the navy at fourteen as well.

He spent the years before the Great War gaining experience. He developed a speciality in navigation and surveying and studied French and Japanese. But his career was held back by a series of incidents that stripped some of the lustre from his obvious emerging talent. In 1901, at the age of twenty-one, he was reprimanded by a court martial for making improper alterations to the log of *Viper*, a fast torpedo-destroyer that was wrecked off the Channel Island of Alderney. Six years later, he was involved in another collision, this time when the cruiser *Flora* ran aground entering Hong Kong's harbour one misty, moonlit evening. Although there were mitigating circumstances—a docking buoy was found to be 450 yards from the location marked on Admiralty charts—Grieve earned a reprimand for poor judgement. In 1911, these errors of propriety and judgement gave way to violence when, aged thirty-one and serving on the battleship *Triumph*, he was court-martialled for hitting an officer and then sending a letter threatening another. This time, the reprimand was severe: Grieve was dismissed from the ship.

His punishment was a chastening experience. He served for a time on a merchant ship, then got a shore job at Devonport Dockyard, on England's south coast. He seems to have worked hard to regain the trust of his superiors, because he was soon back earning their praise. While there, he demonstrated another side to his impulsive character: a willingness to risk his own life to save another.

The incident happened early one September morning in 1913, when Grieve was walking alongside Devonport's north lock just as the battlecruiser *Indefatigable* was passing through. He stopped to watch as sailors attempted to lower a boat over the side of the ship's deck. Suddenly, one of them, a nineteen-year-old former labourer from Belfast named James Reilly, lost his footing and dropped into the cold, deep waters of the lock. Unable to swim, and thrashing around wildly, he began to sink. Immediately, Grieve stripped off his heavy uniform coat, jumped into the lock, and swam with fast, powerful strokes towards the struggling young man. He reached Reilly in seconds. But the terrified sailor grabbed hold of him with such force that he began to drag Grieve under the surface with him. Grieve couldn't free himself from Reilly's panicked embrace—a seemingly instinctive reaction known as the "death grip."

The men who by now were lining *Indefatigable*'s deck and the walls of the lock looked on in dread as Reilly and Grieve fought to keep their heads above the water. It looked like both would soon go down. But Grieve would not submit. Gathering all his strength, he succeeded in subduing Reilly. The reports don't explain how he did this—one of them just mentions "a struggle." Perhaps Grieve managed to peel Reilly's arms from around him and release the death grip that way. Or maybe he had to be a little rougher. A punch in the side of the head or the neck could have been enough to knock the lad out for a moment; even Boy Scouts learned that headbutting a drowning man's nose might daze him enough that he lets go. Whatever had to be resorted to, Reilly released his mad grip long enough for Grieve to get a proper hold of him. Then he pushed the sailor into a lifebuoy that had been thrown in and dragged him through the water to a set of steps in the lock wall, where onlookers hauled him off to the medics. Grieve, who was barely even winded from the whole affair, earned an award from the Royal Humane Society for his bravery that morning.

It was in 1916 that Grieve was appointed the navigator and acting commander of *Campania*, overseeing its aircraft take-off trials with Sopwith the following year. Navigational skill was crucial when the lives of aeroplane pilots relied on the perfect handling of the vessel's every movement. Evidently, Grieve had impressed Hawker and the Sopwith managers. He was also commended by the Admiralty, which remarked on his skill in both navigation and wireless operation, an unusual combination in a single officer. A court of inquiry following the accidental sinking of *Campania* in the Firth of Forth in November 1918 exonerated him, clearing any blemish that might have sat on his record. When, on January 12, 1919, Grieve asked the Admiralty for permission to join the Sopwith team, it was granted.

THE SOPWITH ATLANTIC, AS HAWKER AND GRIEVE'S ONE-OFF AER-
oplane came to be known, had a forty-six-foot wingspan. Its deep, flat-
sided fuselage contained fuel tanks that could hold 350 gallons of petrol,
enough for twenty-two hours of flying at an average speed of 100 miles
per hour. The twin-wheeled undercarriage could be jettisoned into the
sea once the machine had taken off. This would reduce weight and drag
in the air, increasing speed and saving fuel, though it meant that a successful landing would require all of Hawker's skill. A wireless set was
powered by a small, wind-driven propeller that hinged out from the side
of the fuselage.

The most unusual feature of the Sopwith Atlantic could be found on
the top, behind the two seats. The streamlined fairing there was detachable and shaped like an inverted boat. If, for any reason, Hawker had to
bring the machine down onto the waters of the Atlantic, he and Grieve
could unclip the fairing and use it as a lifeboat. It held emergency rations,
a gallon of drinking water, flares to attract attention by day and night,
paddles, an inflatable airbag to provide buoyancy, and an anchor. The
airmen would be wearing rubber life-saving suits designed by the US
Navy, which were fitted with airbags on the chest and at the neck, providing insulation and buoyancy while keeping out the frigid water of the
ocean. Of course, they hoped they would never have to use the boat. But
it provided reassurance that they had at least a chance of survival, should
the worst happen. If the aircraft's single Rolls-Royce Eagle engine should
fail mid-flight, the boat might just save the men's lives. Every aviator
knew that loss of power was a real risk. That spring, the aviator and aircraft constructor Hilda Hewlett informed a reporter, "All the pilots will
tell you that their fate rests with their engines."

The design and construction of the biplane was carried out at the
company's Kingston factory and a facility in nearby Ham and took just
six weeks. Then it was sent by lorry from Kingston to Brooklands, a ten-
mile ride. It was a journey that the Sopwith team, and the residents of
Kingston, had come to know well. Since the Sopwith factory had been

established in 1912, all finished aircraft had left the facility by road, and most were transported to Brooklands for flight testing. The sight of a Daimler flatbed lorry and trailer weaving its way through the narrow Kingston streets, laden with fuselage, wings, and crates containing components, became a familiar sight during the long years of the war, when thousands of aeroplanes left the factory this way.

Final assembly and testing of the Sopwith Atlantic took place at the Brooklands flying village. The most convincing of the test flights was one in which Hawker, accompanied by Grieve in the navigator's seat, piloted the craft at its cruising speed of 100 miles per hour for more than nine hours. This 900-mile flight over southern England covered a distance equal to half that of the transatlantic route. Hawker finished the flight "as fresh as paint," according to one report. Upon landing, he found that two-thirds of the aircraft's fuel remained unused in the tanks.

On March 18, 1919, Sopwith's general manager formally lodged an entry for the *Daily Mail* contest. Then the firm announced to the press that Harry Hawker and Mac Grieve were going to fly the Atlantic.

THE ROUTE SELECTED FOR THE TRANSATLANTIC ATTEMPT started at Newfoundland, the easternmost land in North America, and ended on Ireland's west coast, the shortest distance across the Atlantic. Newfoundland had long held a special place in the history of communication. "In the past Newfoundland has been a part of the great connecting links between the old and new worlds," a newspaper noted at the time. "On the water, under the water, over the water and now the fourth through the air. When the discoverer set out from the other side in the misty past he landed on the shores of Newfoundland, the Atlantic cable came here, Signor Marconi sent his first wireless message from Signal Hill and St. John's is now the starting point of the greatest air flight ever undertaken."

Newfoundland had been an English colony since the early seventeenth century. It became self-governing in the 1850s and secured dominion status in 1907. At about 42,000 square miles, the island is only one-fifth smaller than England and compares in size with the US state of Virginia. The capital city, St. John's, facing east into the Atlantic on the island's Avalon Peninsula, lies on a similar latitude to that of Paris. But Newfoundland forms a severe landscape. A mountain range rises along its western coast, an extension of the Appalachians, and the island is generally hilly. Its coastal cliffs—forming "a brown, bleak and rugged barrier," according to one chronicler—can be tall and forbidding. Deep bays pierce its 6,000-mile-long coastline, some of them pushing ninety miles inland.

The decision to fly the route from west to east was made to take advantage of the prevailing westerly wind, which could rise to about 45 miles per hour at 2,000 feet, making a big difference to speed, fuel consumption, and flight duration. But it meant that the aeroplane, as well as its ground and flight crew, would first have to make the long trip across the Atlantic by ship. Then the aircraft would have to be rebuilt and prepared once more for flight.

On March 20, after two intensive months of design, construction, and testing, the day finally arrived for Harry Hawker to leave his wife and newborn daughter and head for North America. Muriel accompanied her husband from their home in Hook as far as London. Harry would take the train from there to Liverpool, from where the steamship *Digby*, loaded with crates containing the dismantled aircraft, would depart for St. John's. It was one of the hardest journeys the couple would ever make together.

"During the preparations my courage had remained high," Muriel remembered, "but when I went into Harry's room just before we left, and found him crying, I lost heart and broke down entirely. He had been putting a few last things into his bag when his feelings got the better of him. He was always sensitive and soft-hearted, and I knew he was going to be terribly homesick until he got over the other side and had plenty to do. The sight of his grief was too much for me."

Eventually, the pair recovered their composure. They were now running late. Together, they walked out of the house, climbed into the Sunbeam sports car standing outside, and set off. The day was freezing cold, with heavy rain. Muriel sat huddled on the floor, sheltering from the weather. Harry pressed the car quickly through the wet streets, the roar of its engine reverberating from the buildings and traffic as they passed by. Muriel rested her head sorrowfully on her husband's knee as he drove. "I longed to sleep away the next two months," she later admitted.

Only too quickly, the pair reached London's Euston railway station and ran inside. As Harry raced towards the train, climbing aboard with just three minutes to spare, Muriel was left standing on the platform with her thoughts and fears for company. Too soon, the whistle blew. It was all Muriel could do to keep her calm as the train pulled away and slowly wheezed out of the station. Then it was gone, and so was her husband. There was nothing she could do now but wait.

Muriel Hawker went back home to the small villa in Hook, and to baby Pamela, her emotions in turmoil.

===== CHAPTER 7 =====

AS CLOSE AS A CLAM

THE ARRIVAL OF AN AEROPLANE AT ST. JOHN'S RAILWAY STATION
in the early hours of a Sunday morning caused less of a stir than might
have been imagined. Coverage of the transatlantic contest in the local
press had done little to provoke interest in recent days. One paper carried
a report listing passenger arrivals off the steamship *Digby*. After naming
a few notable St. John's politicians who had been on the ship, the paper
added, almost as an afterthought, "and a couple of airmen to fly across
the Atlantic." One local reporter did take an interest in the arrival of
Harry Hawker and Mac Grieve, however. Joey Smallwood would become
the premier of Newfoundland when the dominion joined the Canadian
federation in 1949. Thirty years before that, he was an eighteen-year-old
cub reporter on the island's *Evening Telegram* newspaper, and always
searching for an angle that would boost his profile among the seasoned
newspapermen of Newfoundland. When he heard that the Sopwith team
was coming to the island, he resolved to be the first person they met
when they pulled into town.

It was a minor miracle that Hawker's party made it to St. John's after
all. *Digby* was 120 miles off Newfoundland when it came upon an impen-
etrable pack of Arctic ice which had been massing along the Avalon Peninsula's eastern coastline for weeks. The ship was built with a specially
strengthened hull to withstand impacts from the floating sea ice often

St. John's railway station in about 1900.

encountered on the Newfoundland route. But this was the worst the coast had experienced for sixty years. Nothing could get through.

The Sopwith aeroplane had been packed into two large wooden crates, each the size and shape of a railway carriage. They were thirty feet long and weighed over five tons apiece. Between them, they held the fuselage, wings, engine, undercarriage, and fittings. Several smaller cases stored in the ship's hold carried parts, tools, petrol, engine oil, and other provisions. It should have been a relatively straightforward offloading job onto the dockside at St. John's. Instead, *Digby* was forced by the ice field to divert to Placentia Bay, off the peninsula's western coast. There, in the middle of the bay, it drew up alongside the postal steamer *Portia*, and the entire Sopwith consignment, as well as the men themselves, were transferred from one ship to the other. Then *Portia* steamed up the bay to the town of Placentia, where its precious cargo of crates and passengers was delivered, leaving *Digby* to continue on its way.

The next stage of the Sopwith saga was a sixty-mile rail journey across the peninsula. The two giant crates were loaded onto flat railcars, and the rest of the cargo was taken into freight wagons. After the cars were hitched to the back of the train, railway officials discovered that it was now too heavy to be hauled all the way to St. John's by its single locomotive. So, once the consignment reached a railway junction halfway across, the cars carrying Hawker and Grieve's shipment were detached and left temporarily in a spur alongside the main line, with the Sopwith men keeping an anxious watch on their cargo. It was late in the evening when the oversized crates and their guardians were retrieved by the returning locomotive and hauled the rest of the way. The special train finally pulled into St. John's station at five minutes to three on the morning of Sunday, March 30.

The party was greeted at the station by Joey Smallwood, who escorted the men by car to the Cochrane, a thirty-two-room hotel on a hillside looking down towards St. John's Harbour. Accompanying Smallwood was a Sopwith employee named Montague Fenn, a thirty-year-old former engineer and army captain who had been Sopwith's field agent in Paris. Now he was managing the transatlantic flight operation. He had been working in St. John's with a small advance party since early March, charged with finding suitable land for an aerodrome. What appeared to be an almost insurmountable challenge when looking at paper maps proved even more difficult on the ground. Harry Hawker later said that Newfoundland was "the last place in which one would look for spacious landing grounds," adding, "if anything the maps seemed to flatter the country."

Smooth and level ground, firm enough to take the weight of a heavily laden aeroplane, was almost impossible to find. The requirement for the site to be located close to the port of St. John's, where it had been expected the aeroplane would arrive, further limited Fenn's options. And the orientation of the aerodrome was crucial: the direction of the prevailing wind could mean the difference between success or crushing failure. A further challenge had become apparent to Fenn and his crew as soon as they stepped off the steamer that brought them to St. John's. Much of Newfoundland was still blanketed in snow.

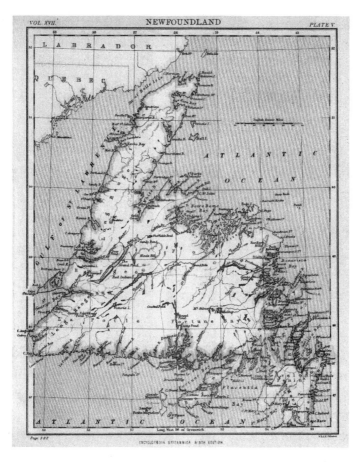

Map of Newfoundland. St. John's, the Avalon Peninsula, and Placentia Bay are at the bottom right.

"Captain Fenn certainly got the very best place available," Hawker later wrote. Fenn had selected a patch of land at Glendenning's Farm, sited on a plateau at Mount Pearl, about six miles southwest of St. John's. The L-shaped plot measured about 400 yards along its longer arm and 200 yards along the shorter. At the elbow of the site sat the steeply inclined faces of a 200-foot hill. Trees skirted the aerodrome, part of which was on a slope looking east to the harbour and, beyond, to the Atlantic. Much of the ground was soft and boggy, with no drainage to carry away melted snow. At 450 feet above sea level, the site was a steep climb from the city's

Charles Lester's horses and men dragging a Sopwith crate up to Mount Pearl.

railway station, and its elevation and exposure made it susceptible to gusty and unpredictable winds. But aerodrome-quality land was in short supply. It was the best Fenn had been able to find.

After getting a quick briefing from Fenn, and doubtless answering a few of Smallwood's more urgent questions for the *Evening Telegram*, Hawker, Grieve, and the other Sopwith arrivals snatched a couple of hours' sleep at the Cochrane. Then it was time to get up and get started. They returned to the railway station, where they met Charles Lester, a fifty-seven-year-old trucker whom Fenn had contracted to transfer the crated aeroplane to Mount Pearl. Lester brought twenty men, ten horses, and his largest truck to carry out the job. But it wasn't going to be easy. The weather that morning was atrocious.

Lester and his crew manhandled the heavier of the two crates onto the four-wheeled truck and made a start. Getting the vehicle through the streets of St. John's was straightforward. The challenge came when the men had to leave the roads behind and haul the truck up rutted, sodden, unmade tracks, dotted with trees and scrub, towards Mount Pearl. Rain sheeted down all morning, rendering parts of the route a quagmire. Despite the profusion of horses and men, the cargo quickly became mired in the saturated mud and drifts of snow. In places, the wheels of the truck sank to their tops in the soft ground. Several times, the Sopwith crate and the flagging horses became stuck, requiring leverage from heavy timber props, combined with brute force and a lot of shovelling, to get them freed. At one point the entire crate almost tipped over onto its side. Everybody was relieved when it was finally offloaded at the Mount Pearl aerodrome. The next morning, the gang returned to the railway station to fetch the second crate. Though this was a little lighter, the team struggled just as much. The men tried to remain cheerful, but it was a hard job.

It had been obvious to Hawker and Grieve as soon as they arrived at Mount Pearl that the conditions were no good for flying—at least, not right away. A newspaper reporter at St. John's described "snow and knee-deep mud," observing that the snow was falling daily and melting into the ground. But at least the team would have somewhere comfortable to work while they waited for better weather. Montague Fenn had arranged for the construction of a large wooden hangar and machine shop. It measured fifty-five feet wide by fifty feet deep, with a height of thirty feet, and was large enough to accommodate the entire assembled aeroplane. The lofty apex roof was supported by a lattice of stout timber trusses. Timber buttresses supported the walls from the outside. The building's front was open, shuttered by sliding wooden panels that could be closed to keep out wind and snow, or removed and laid in front of the hangar to provide a smooth floor when the aircraft was pushed out. Ground stretching 100 feet in front of the building was covered with a layer of stone aggregate, which filled the worst of the ruts and holes in the boggy ground. The railway carriage–sized crates that had proved so difficult to transport, emptied of their contents, served alongside the hangar as a site office and meeting room.

The Sopwith team quickly got to work assembling the aeroplane. They hoped to make their getaway just a fortnight later, when a full moon would light their way. But the unpredictable winds experienced at Mount Pearl concerned Hawker. Then, heavy snowfall on April 6, followed by

The Mount Pearl hangar and site office.

twelve hours of continuous rain, turned the already sodden aerodrome into a mud bath, with deep drifts of fallen snow. Any hopes of a quick departure were dashed. It was the last thing the Sopwith team needed. Hawker later recalled, "We just wanted to get the thing over and done with as soon as possible. I'm a little afraid that sometimes the enforced idleness got on our tempers."

For eighteen-year-old reporter Joey Smallwood, meeting the aviators who set up their temporary home on Newfoundland left a deep impression. What Smallwood recalled as much as their mechanical and navigational skills was their character. He wasn't yet a hard-bitten newspaperman, and certainly not the seasoned politician he would become over the decades that followed. He noticed how interviewees made him feel. "Hawker," he later wrote, "was as fearless as he was competent." But Smallwood found him to be "a peculiarly reticent man, rarely opening his mouth, and then only to drop a dry remark." He went on to say, "It has been observed, however—and perhaps with some reason—that his personality was a most unlovable one." Grieve, described by Smallwood as "a tall, spare, broadshouldered Scotchman," was the opposite. "Of the two," the teen reported, "he was the most likeable and the most liked." It was certainly true that Mac Grieve was popular among those who got to know him. What Smallwood hadn't grasped was Hawker's mercurial nature. He could indeed be reticent when he was absorbed in his work or struggling with a problem. But then his mood would switch: once the challenge was overcome, he regained his energy and the charm returned, and everyone around him remembered why they loved him so much.

As they waited for the weather to improve, the Sopwith team completed the aircraft's assembly and tested its engine and wireless equipment. Hawker and Grieve experimented with the insulated life-saving suits and the fuselage lifeboat—Smallwood said it was "more like a very tiny and fragile canoe"—on a nearby duck-pond.

Then, on April 10, a frost hardened the aerodrome's mud a little. That evening, Hawker and Grieve finally got the aeroplane off the ground, making a successful test flight over St. John's and its surroundings that lasted fifty minutes. It was the first aeroplane ever to fly in Newfoundland. In Hawker's accomplished hands the machine reached heights of 3,000 feet and speeds of 100 miles per hour. The streets of the city were crammed with spectators staring into the sky, amazed by what they were witnessing.

AS HAWKER AND GRIEVE SETTLED INTO THEIR ARMCHAIRS AT THE Cochrane Hotel that evening, there must have been a sense of satisfaction and anticipation in their minds. Despite the challenging start to their Newfoundland adventure, with the Arctic ice closing the harbour, and the risky ship-to-ship transfer in the sea off Placentia, and the boggy cart tracks up to Mount Pearl, and the rutted and pockmarked flying field, and the apparently interminable snow and rain, everything now seemed set. The airmen had a functioning aerodrome. They had an aircraft they now knew could take off from it. And they could feel the goodwill of a dominion urging them on to success. Nonetheless, they kept the local reporters at arm's length about when they planned to make their big departure. "Hawker and Mackenzie-Grieve are not communicative," wrote the *St. John's Daily Star* after the test flight, "and have not taken anyone in their confidence as to their intentions in this respect." The paper later said Hawker was "as close as a clam." The men would have to get away soon, though, if they were to win the contest. All they needed now was a favourable weather report for the mid-Atlantic and they would be ready to go.

The next morning, news reached Mount Pearl from St. John's Harbour. The steamship *Sachem*, from Liverpool, had just docked. On board was Harry Hawker's old Brooklands friend, Fred Raynham. With him was a disassembled aeroplane, a team of expert mechanics from Woking, and Charles Morgan, the former naval midshipman whom Jack Alcock had taught to fly four years earlier. The Martinsyde company had arrived, and the Sopwith team now had competition.

CHAPTER 8

ALL SORTS OF STUNTS

BY THE TIME HE ATTENDED A MEDICAL EXAM IN NOVEMBER 1918, Charles Morgan was in the grip of a condition then described as neurasthenia. It could mean different things in different contexts. In Morgan's case, it described a collection of symptoms resulting from traumatic events he had experienced during the war. It was akin to shell shock. The leading symptom—one which had been dogging him for many months by the time of his consultation—was depression. It was a debilitating condition for the twenty-nine-year-old sailor-turned-airman. It made it difficult for him to concentrate for any length of time and his memory was poor. He tired easily, which made him irritable. During the day, he endured abdominal pain and headaches over his left eye, and his hands shook when he extended his arms. Worse was to come each evening after he slipped into broken slumber. Nightmares invaded his sleep, the same terrifying experience each time. Charles Morgan dreamed, night after night, that he was being blown up.

Morgan joined the Royal Navy as a cadet in 1905, at the age of sixteen. After successfully completing his cadetship, he transferred to the merchant service and the Royal Naval Reserve, serving in the West Indies, the Indian Ocean, and the Persian Gulf. In 1911, he married a nineteen-year-old woman from Brisbane named Phyllis Anderson. They had their first child in July 1914, just before war was declared. Soon afterward, Morgan was posted to the British hospital ship *Rohilla* as its navigating officer.

In the early hours of Friday, October 30, *Rohilla*, a large, fast steamer with powerful twin engines, was travelling from Scotland to Dunkirk. It had about 230 nurses, doctors, officers, and crew aboard, and its mission was to evacuate wounded soldiers from the battlegrounds of northern France and bring them back to Britain for treatment. The weather in the North Sea off Britain's east coast that night was dreadful. Heavy, driving rain drenched the vessel as it steamed south. As it passed North Yorkshire, about one-third of the way through its intended journey, it experienced violent gale-force winds building mountainous waves that pounded the coastline. Navigation beacons and sounders that marked treacherous rocks and reefs had been switched off owing to wartime restrictions.

As the weather worsened and his heavily laden ship began to roll in the storm, the vessel's fifty-year-old Scottish captain, David Neilson, adjusted course. He wanted to skirt around a minefield that had been laid to protect the coast from attack. Standing on the deck to supervise the ship's progress, he observed vessels passing in both directions, giving him confidence that his course would take them clear of the mines. But without being able to sight navigational lights, he was unable to see that the weather was driving the ship farther inshore than he thought. Just before four o'clock that morning, a local coastguard at Whitby began signalling with a Morse light to warn the ship of impending danger. But *Rohilla* was too far away for its lookouts to decipher the message. A few minutes later, the vessel drove hard onto a reef, 400 yards off the Whitby shore.

As the ship struck the rocks with a dreadful grating noise and juddered to a halt, Neilson shouted that they had hit a mine.

"No sooner had we struck than she began to break up like matchwood," recalled one survivor. With a rising tide and the ferocious storm, the vessel quickly started to become submerged, with heavy waves lashing the decks. All but one of the vessel's lifeboats were destroyed by the water. The ship's officers repeatedly fired distress flares into the oil-black

Rescuers at the wreck of *Rohilla*.

sky, hoping they would help those gathering on the shore to locate the steamer and send assistance.

The first three bodies washed onto the darkened shoreline north of Whitby at 5 a.m. The violence of the storm had stripped the men's corpses almost naked.

As day broke over the scene, Whitby's residents crowding the shore began to understand the scale of the unfolding disaster. The crew of an onshore lifeboat made two journeys through the crashing waves to the wreck. They rescued a total of thirty-five people, including a forty-four-year-old stewardess who had survived the *Titanic* disaster two years earlier. But the men were forced to abandon further rescue attempts until the storm abated, leaving the other survivors crammed into the ship's bridge or hanging onto rigging attached to the masts.

By 8.30 a.m., the rear portion of the vessel had started to break away from the main body of the ship. It was held only by the propeller shafts. In the afternoon, it broke off entirely. In view of the appalled onlookers lining Whitby's shore, eleven people clinging to the stern were washed away and drowned as the wreckage crashed down into the sea.

As the day wore on, some, desperate, tried to swim through the howling gale for the shore. A few made it, pulled from the water by Whitby residents who formed human chains from the beach out into the foaming shallows. Many more drowned in the attempt. One swimmer made it almost to the shore and the hands of a rescuer when his head struck a rock. Seconds from salvation, the man died in the breakers.

Efforts to launch rescue lines to the ship using rockets failed. Several people died from exposure that afternoon.

As night fell, the front of the vessel fractured from the central portion and sank fast. Hopes began to fade for the remaining crew of fifty, including Charles Morgan, who were left huddled in the bridge and on the rigging as the storm continued to assault the ship and the waters around it.

By noon on Saturday, thirty-two hours after the vessel first struck the rocks, more bodies had washed ashore. Makeshift rafts fashioned from doors and shutters were launched from the ship, but almost all those who attempted to reach the shore using them perished during the journey. That afternoon, the ferocity of the storm began to subside a little. By Sunday morning, many of those still alive on the wreckage of the ship had been rescued by lifeboats or managed to swim ashore. One man was carried up the beach with blood coursing from deep lacerations to his feet and legs. He had slipped on the rigging and fallen. Those who were last to make it off the ship had endured fifty hours on the wreck with little water, no food, and no fire to keep them warm. Morgan was one of them.

Somehow, *Rohilla*'s flag found its way from the ship to the shore. It might have washed up with wreckage or been carried by one of the survivors. Perhaps it was even brought by Morgan himself. A few weeks after the disaster, once he had recovered from the double pneumonia he contracted on the wreck, Morgan presented the flag to his local church of St. Andrew's, in the Hampshire village of Tiptoe. His wife had offered prayers there on the evening of October 30. She had taken their three-month-old daughter with her. A church magazine later observed that "those who look at it should regard it not merely as an interesting relic of a naval misfortune, but also as a grateful offering of thanksgiving for a merciful deliverance vouchsafed to anxious prayer."

In total, eighty-three people lost their lives in the grounding of *Rohilla*. Some bodies were never recovered. The ship's captain, David Neilson, told the coroner's inquest he believed his ship hit a mine outside the reef. The blast felt by those on board—"a sudden crash," as one of the surgeons reported to the inquest—was likely to have been a consequence of the speed with which the vessel struck the rocks. The explosion of rockets used to fire rescue lines to the ship in the aftermath of the grounding would have added to the terror of those left on the rigging amid a raging storm. The volley of distress flares sent up by the ship's crew—a repeated percussive thump—contributed to the sense that the vessel was under attack. It is easy to comprehend how some on board must have believed that the ship was being blown up: that *they* were being blown up. And just as easy to appreciate that the memory of those terrible fifty hours could haunt Charles Morgan's mind, and his body, and his dreams, years later.

AFTER RECOVERING PHYSICALLY FROM HIS ORDEAL IN THE WRECK of *Rohilla*, Morgan secured a transfer to the Royal Naval Air Service. He was sent to Jack Alcock's flying school at Eastchurch and passed his qualifying tests in March 1915. But bad fortune followed him there. Surviving records of Morgan's wartime experiences are sketchy in places. One report suggests that he was involved in a flying accident at Eastchurch in which the wings of his aeroplane collapsed mid-flight. The machine fell 1,200 feet to the ground. In the resulting crash, Morgan sustained injuries to his left leg as well as suffering a cut over his left eye. Then, on May 15, Morgan got shot in the right foot. By all accounts, it was accidental. The gun was reportedly fired by Robert Hilton Jones, a flight lieutenant later described as possessing "a certain wildness of character." The bullet was removed from Morgan's foot, but the wound worsened over the following months. In February 1916, Morgan's right leg was amputated just below the knee.

With an ill-fitting prosthetic, Morgan spent the rest of the war in ground roles. In 1917, he took over the compass facility at Brooklands

Aircraft Acceptance Park, one of the sites where aeroplanes were given their final tests before being issued with navigational instruments and sent to the fighting fronts. But his physical and mental health continued to deteriorate. In May 1918, soon after the birth of his and Phyllis's second child, he was invalided out of the service.

The medical exam that took place in November that year, which noted Morgan's neurasthenia, was part of the process needed to secure a pension owing to wounds and illness sustained in the war. Sick, lame, and unfit for service on the ships or aircraft he had trained for, thirty-year-old Morgan, like many others emerging from the war, was in a bind. He had a wife and two young children, and he would need more than a meagre wound pension to sustain the family. Employment opportunities for men broken by the conflict were scarce. Morgan was an outgoing, animated character, talkative to the point of garrulousness. After leaving the air force, he needed to expend his restless energy. He took on the chair of the Surrey division of the Comrades of the Great War, a veterans' organization that would later form part of the British Legion. He also ran its Byfleet branch, positioned midway between Brooklands and Woking, where Martinsyde's extensive factory was located. But these were unpaid positions. Morgan needed a job.

In August, he struck it lucky, or so he thought. He got a job as a political organizer for the British Commonwealth Union, a business group lobbying Parliament for protective tariffs. He was given the Birmingham patch and attacked his new duties with characteristic gusto. To begin with, he seemed to be performing well. A couple of weeks into the role, his first report to the central office was received with satisfaction. Matters quickly soured. Over the next few weeks, the group's managers decided he wasn't up to the job. In early November, with just over a fortnight left on his three-month probationary period, Morgan was told that his services were no longer required. It was just a few days after he had received the diagnosis of his neurasthenia, with its debilitating physical symptoms, its neurological impairment, and the mental conditions that followed the traumas he had experienced early in the war. As he tendered his resignation to the union's executives, Morgan explained to them something of his personal circumstances. They recognized the difficulty of his situation, and agreed to pay him up to the end of his probation, an extra £25 in salary that he needed badly. But that was all.

The Armistice had just been signed. War was over, and countless veterans—many, like Morgan, with disabilities—were queueing at the labour exchanges in search of work. It was the worst possible time for Morgan to find himself unemployed again. When word reached him that the Martinsyde company was planning to fly the Atlantic, he must have wondered what he could possibly lose by trying to get involved.

WHEN THE *DAILY MAIL* FIRST ANNOUNCED ITS TRANSATLANTIC contest in 1913, Martinsyde was the only British company that made serious plans to compete. But the untimely death of its star pilot, Gustav Hamel, set the company's progress back badly. Before it could recover, the outbreak of war intervened, and the contest was put on hold. After the war ended, Martinsyde was hit as badly as Sopwith had been. Military orders for 1,500 of its most recent type of aeroplane were summarily cancelled. The firm managed to stay afloat in the months following the Armistice, with some aircraft production continuing and export orders energetically pursued. But, like Sopwith, it badly needed the publicity that a successful transatlantic flight would bring.

During the war years, Fred Raynham had tested hundreds of Martinsyde's aircraft at Brooklands. As the firm's chief test pilot, his contribution to the company's success had been considerable. "Although young in years," a newspaper later reported, "he is an 'old bird' who knows all the tricks of the trade and the air." When Martinsyde began to plan its post-war transatlantic flight, it must have been obvious that twenty-five-year-old Raynham would be the person to pilot the aircraft. But who would be the navigator?

There can be little doubt that Raynham and other figures at Martinsyde would first have met Charles Morgan when he ran the compass department at Brooklands. Martinsyde staff also attended sporting fundraisers that Morgan organized for his veterans' group in nearby Byfleet.

As Martinsyde looked around for a navigator, they must have noted that Morgan had the necessary technical skills, both as a navigator and an aviator. Those who had taken part in his veterans' fundraisers knew him as a capable and energetic organizer as well. And, like the ex-servicemen he spent his days trying to help, he was in need. So, they gave him the job.

The flight crew had been picked. Now they would have to build an aeroplane and find somewhere in Newfoundland to fly it from.

MARTINSYDE'S TRANSATLANTIC AIRCRAFT WAS BASED ON THE company's F.4 biplane, the fastest aeroplane in the world at the closing stages of the war, and one which showed great promise. The competition version would be a little larger than a standard F.4, although, with a wingspan of forty-three feet, it would be smaller than the Sopwith Atlantic. Morgan, the navigator, would be seated up front, with pilot Raynham sitting behind. It had dual controls, meaning Morgan could pilot the craft should the need arise. Powering it was a single Rolls-Royce Falcon engine. This was like the Eagle engine fitted to Hawker's craft, and just as reliable. But at 275 horsepower, 100 lower than the Eagle, it had less grunt.

There would be no special life-saving devices such as lifeboats on Martinsyde's aeroplane. It was simple, sturdy, and stripped-down. Nor would its two-person crew wear any special clothing over their ordinary flying suits, besides a life-vest each. While the Sopwith team would be outfitted in rubber flotation suits, which might keep them alive for a while if they had to ditch in the ocean, Martinsyde's airmen scorned such measures. Morgan said to reporters, "I am afraid these life saving gadgets will be of little use. For myself, I have decided that I may as well take one deep breath if we strike the water. We will be a small speck on a big ocean out there." Raynham said they proposed to "cross the Atlantic, not fall into it," adding that "if a fellow is going to drop in he may as well go down as float around and freeze."

With work started on the aeroplane, the firm needed to reconnoitre possible aerodrome sites in Newfoundland. As Montague Fenn of the

Sopwith company would later discover, this was not a task that could be carried out using maps alone. It would demand on-the-ground knowledge of the challenging topography and climate of the region. An understanding of wind direction, the feasibility of access routes, the capabilities of local trucking contractors, and a host of other practical details would have to be established. Then, once a site had been identified, negotiation and deal-making would be required to secure a lease. Moreover, the person carrying out this important reconnoitre would be building a network of supporters ready to assist the Martinsyde contestants when they arrived on the island in March or April, ready to compete.

The task seemed tailor-made for Morgan. He was gregarious, energetic, and eager to prove himself capable. On January 30, 1919, buoyed by his new position and responsibilities, Morgan set out for St. John's.

THE CROSSING, DELAYED BY THE ICE BEGINNING TO CROWD THE Avalon coastline, took eight days. At some point during the voyage, Morgan caught influenza, so, when he reached St. John's, he was taken straight to the military hospital to recover. Word of his arrival reached the ears of Joey Smallwood at the *Evening Telegram*. On February 14, Smallwood visited Morgan in his hospital bed, excited to scoop an interview with a transatlantic flight contestant. Morgan impressed Smallwood, who found him to be "open and free in disposition." He added that Morgan "would discuss the machine, her chances, and anything connected with it, without the slightest hesitation, and even seemed to like doing so." Another reporter would describe Morgan as "one of those English sports who immediately wins his way into the hearts of those he meets. He has a jovial disposition, always sees the funny side of life and it is not any wonder that he has lots of friends. . . . He is of the bull-dog class—one of those who never knows defeat. His close friends in St. John's simply swear by him, as a better mixer they say it would be difficult to find anywhere."

During his interview, Smallwood advised Morgan to look for aerodrome sites in Pleasantville, north of Quidi Vidi Lake, about a mile or

two north-east of the city centre. Morgan got similar advice from leading figures in the dominion: the governor, the minister of militia, and
the senior naval officer. Ten days after meeting Smallwood, Morgan had
recovered enough to be discharged from hospital, and he headed straight
for Pleasantville. He soon found what he was looking for: a spot of rea-
sonably level ground in the form of an old cricket pitch on the north
shore of the lake. He secured agreements to lease the site, returned to St.
John's, and finalized his affairs. Then he set off back to England, leaving
behind a small army of fans who were eager to see him return and have
a go at the prize.

Morgan got back to Martinsyde's Woking factory to discover that in
the seven weeks he had been away, the aeroplane had been completed and
was now ready for flight and engine tests. It performed superbly in every
trial. The longest test flight took place on Sunday, March 23. With both
Raynham and Morgan on board, Raynham piloted the craft to Southampton and back five times, remaining in the air for ten hours. Afterward, the Rolls-Royce engine was stripped down to its component parts
and inspected minutely for wear or damage. It was in perfect condition,
as if it was fresh from the factory. Both men enjoyed the flight, and the
aircraft itself performed admirably, whatever altitude it was flown at and
however fast Raynham pushed it. Everyone agreed that the test had been
a comprehensive success.

The following day, a *Times* correspondent was invited to visit the factory and see the machine first-hand as it was disassembled for transit to
the Liverpool docks. He was told that the aircraft had been christened
the Raymor, a contraction of the names of its two crew, Raynham and
Morgan. He was shown the 370-gallon aluminium fuel tank, holding
sufficient petrol to carry the aircraft at 100 miles per hour for twenty-
five hours. He inspected the glazed windows and adjustable roof enclos-
ing Morgan's seat, protecting him against the elements as he made his
observations and prepared calculations at a small chart table equipped
with miniature maps. The quarters, though confined, looked comfortable and secure.

Morgan was on hand for an interview. "He is enthusiastic for the ven-

Quidi Vidi Lake, about 1920.

ture," noted the reporter, "and told me to-day that he has been greatly impressed by the keenness of people in Newfoundland on cross-Atlantic flight. The start, of course, is to be made from that side, and everything that could be done to help over there has been done by the many friends he has made in the Colony." But Morgan sounded a note of caution. In another interview, he said, "The conditions for flying in Newfoundland at this time of the year are generally bad on account of the deep snow and boisterous weather, and we may experience some difficulty in getting away." This comment would prove prophetic.

Now they were as ready as they would ever be. On March 26, Martinsyde lodged its formal entry into the *Daily Mail* contest. The following day, as the Sopwith team on *Digby* was facing the ice field blocking St. John's, Martinsyde's biplane, packed in a single giant crate, began its slow journey by road from Woking to Liverpool to meet the steamship *Sachem*. That evening, Morgan was guest of honour at a farewell dinner given in Woking by Helmuth Martin and George Handasyde. In an animated after-dinner speech, Morgan told the audience how well the ground had been prepared at Newfoundland. He said that even though the Sopwith team had got out there first, he believed the Raymor

would make it back to British shores before them. At this, the happy
crowd cheered.

THE *NEW YORK TIMES* REPORTER STATIONED AT ST. JOHN'S WROTE,
"As if luck were deliberately favoring the tardy arrivals, the *Sachem* slid
into port without the slightest trouble." The steamship arrived at 10 a.m.
on Friday, April 11 after an eight-day voyage from Liverpool. Its arrival
could not have been more different in character from the tribulations
experienced by *Digby* a fortnight earlier. A stiff wind from the south-west
had broken up the packed Arctic ice and blown it far out to sea, leaving
the port clear for the steamship's arrival. Before the first passengers had
even completed their customs paperwork and passed from the pier into
the city, the Martinsyde crate had been connected by stout lines to the
deck cranes. By mid-afternoon, it had been hoisted out onto the dockside, and before long, it was packed securely onto a waiting wagon ready
to be carted to Pleasantville the following morning.

Martinsyde's consignment included a large canvas hangar, a design
that had seen extensive use at the British front during the war, and it
would take a mere hour to erect if the weather allowed it. Martinsyde
therefore had no need to build a large wooden structure, as Sopwith had
done; the newcomers would gain days on their rivals just on this matter
alone. And the ground at Pleasantville, resting on rock and well drained,
was turning out to be firmer in this wet Newfoundland weather than at
Mount Pearl. It seemed as if the scouting work carried out by Charles
Morgan back in February was bearing better fruit than Montague Fenn's
activities a month later.

While the cargo was unloaded from the steamship, Raynham, Morgan, and the rest of the Martinsyde group went to the Cochrane Hotel,
where they had booked rooms. Hawker and Grieve were out, working
at Mount Pearl. The Martinsyde men left soon after checking in their
belongings and headed for Pleasantville. The Brooklands reunion would
have to wait.

Newspaper reporters watching the transatlantic contest observed a mood of frustration rising among the Sopwith team once news of Martinsyde's arrival reached them. A special publicity dress rehearsal had been planned that afternoon at Mount Pearl. With newsreel cameras recording the scene, the governor of Newfoundland, together with other dominion officials, was to say a formal farewell. Then he would hand over a package of letters for the contestants to carry—including one to King George V—and watch Hawker and Grieve take off on a second trial flight. The dignitaries, resplendent in tall fur hats, crowded into the aerodrome. Hawker described the aeroplane and answered questions. At 4 p.m. the machine was pushed out of the wooden hangar. To great acclaim, the propeller was swung, and the Rolls-Royce engine growled to life. But after a twenty-minute test, Hawker shut the engine off and retreated into the hangar, refusing to take the machine off the ground. He later blamed the weather. The field was boggier than the previous day and fresh snowfall was forecast. But his reluctance to entertain the delegation was interpreted by onlookers as ill-mannered, if not downright rude. The *New York Times* reporter on the scene said Hawker rivalled "the operatic prima donna." As officials, mortified, tried to persuade him to change his mind, "he only grew sulkier," the reporter noted. The governor and his party finally left the aerodrome, dissatisfied.

The reality was, Hawker was panicking. He was already under pressure to make the mid-April window of opportunity with the moon and the weather. Now he faced competition from a well-prepared and fast-working Martinsyde team as well. Raynham's men had a better aerodrome and a simpler aircraft. The last thing Hawker could afford was to risk a second test flight with such poor conditions underfoot. Any damage coming in to land, even in a lightly loaded machine, would take days to repair, or possibly longer, and they were days he no longer had. Now, he and Grieve would be fighting against Raynham and Morgan hour by hour to be first away.

Throughout Saturday the 12th, both teams worked at a fast clip. For Raynham and Morgan, the day was spent supervising the team carting their crate to Pleasantville. High winds prevented the Martinsyde engineers from erecting their canvas hangar. By the evening, although the crated aircraft had arrived at the aerodrome, it was obvious that the newcomers would not be able to get away until Monday at the earliest.

At Mount Pearl, a gang of Boy Scouts shovelled broken rocks to form a makeshift runway from the wooden hangar the length of the aerodrome. Labourers filled ruts and pockets and pushed rollers over the hummocks that pocked the site, trying to flatten it for the heavily laden take-off. Inside the hangar, mechanics, fitters, and engineers performed an exhaustive roster of final tests and inspections. Petrol was filtered obsessively through chamois cloths before being loaded into the aircraft's capacious tanks. The little fuselage lifeboat, all six feet of it, was packed with provisions and equipment and clipped carefully to the aeroplane. At one point, the signal pistols that could fire bright red flares to attract attention were tested. The signals soared into the sky, but within two seconds they burned out. The Sopwith technicians, realizing they would be next to useless over the vast ocean, packed them for Hawker and Grieve to take anyway. Throughout the day, Hawker paced nervously around the site. Sometimes he would inspect the adjustment of a particular mechanism or try out newly connected controls. Most of the time he could do nothing, and merely looked on as the men worked on their tasks.

At mid-afternoon, a visitor arrived at Mount Pearl. It was Fred Raynham. "The two lean, thin-faced, clear-eyed young men grinned broadly at each other," wrote the *New York Times* reporter, "and then turned calmly to a discussion of the techniques of their game." The pair ribbed each other good-naturedly about the relative merits of the two aerodromes. Then Raynham cast his eye over Hawker's aeroplane. "My boat is quite a little fellow beside yours," he said. After chatting for a few more minutes about the weather, the pair parted. "Good luck, see you again," was all Raynham said to his old friend as he walked away.

At 4 p.m. a strong, cold wind that had been buffeting the aerodrome all day picked up. It began to push rain over the plateau, which started softening the already soggy ground. At 5.30 thicker rain clouds began to encircle Mount Pearl. With the aeroplane now fuelled up and ready to fly, Hawker and Grieve gathered in the lofty wooden hangar with the rest of the Sopwith team. Now was the moment to make the decision. They could take off now and beat Martinsyde into the air. But the weather was worsening. Moreover, it would be dark when they reached the other side of the Atlantic. Having jettisoned the undercarriage, Hawker would already struggle to land the craft safely. Doing so blind would, in the words of the *New York Times*, be "practically suicide."

After a long conference in the hangar, Hawker emerged with the team's decision. "The machine was in readiness, the aviators had only to jump in, but the weather man was in bad humor," reported the *St. John's Daily Star*. If the Sopwith team had completed work on the aeroplane earlier in the day, they might have got away. But now the closing storm was against them. The airmen were not suicidal; there would be no attempt that evening.

It rained all night over Mount Pearl. On Sunday morning, the aerodrome was, once more, a quagmire. As the day advanced under heavy skies, a thick fog descended as well. Despite the mud and murk that prevented his departure, Hawker seemed to be in a better mood now that his aeroplane was ready. While the Sopwith mechanics continued fine-tuning the machine, Hawker and Grieve drove to the Martinsyde airfield in Pleasantville to see how the competition was faring.

Martinsyde had worked fast. They had got their canvas hangar up and then assembled the aeroplane's fuselage. The wings were ready to be attached. They expected to be ready to depart the following day, though that depended, as with the Sopwith team, on favourable weather.

A few days before Hawker and Grieve had arrived, a small group of British meteorologists landed at St. John's to study wind and weather patterns on the island and over the Atlantic route. Stations had been established at Lisbon, on São Miguel island in the Azores, and on a British battleship stationed in the ocean between the Azores and St. John's. A special coordinating bureau receiving their reports had been set up on Duckworth Street, in downtown St. John's, under the control of a young Royal Air Force meteorologist named Lawrance Clements. Clements had bad news for the two teams that weekend. A depression over the Atlantic meant the weather conditions were heading from bad to worse. One newspaper said, "The wind is doing all sorts of stunts. Every two or three minutes there is a change and one never knows what is coming next." With such unpredictable weather along the route, Hawker, at 4 p.m., announced that his start would be postponed once more.

For the next three days, the Martinsyde team laboured over the assembly of the Raymor. The Sopwith team, with little left to occupy their time, waited impatiently for the weather to turn. But the Atlantic storms continued to rage, and rain continued to fall at Mount Pearl, keeping the ground soft and creating pools and rivulets in the uneven surface of the aerodrome. Each day, knowing that the delay favoured the Martinsyde team, Hawker had to postpone his take-off. Then, on Thursday, April 17, at around midday, Fred Raynham and Charles Morgan made it into the air over Pleasantville.

It took just fifty yards of taxiing along the bumpy field before the stocky Martinsyde machine lifted off the ground. "The biplane fairly leaped into the air," observed the *New York Times*. "It was a beautiful rise," agreed the *Evening Telegram*. "The noise of the engine—sounding like five hundred brass drums going in full blast—could be clearly heard by the people gazing skywards." Among those watching at the Pleasantville aerodrome were Hawker and Grieve, who chatted with Raynham and Morgan as the Martinsyde mechanics made final adjustments to the aircraft before starting its engine.

It was a flawless test flight. For more than thirty minutes, Raynham circled, ascended, descended, and made high-speed straight runs along the one-mile length of Quidi Vidi Lake. He also took photographs of the scenery below. By the time he brought the machine down in a perfect landing back at the aerodrome, he had covered seventy miles, thrilling the St. John's crowd in the process. It was only the second aeroplane flight they had ever witnessed in the dominion.

It was a disappointment for Hawker and Grieve. They had now lost the advantage of arriving at Newfoundland earlier than Raynham and Morgan. But the two pilots, as old friends, knew that each was a worthy opponent of the other. They had worked together for years and had nothing but respect for each other. They were two of the finest, most experienced, most capable pilots in the world, and they were in good shape.

Conception Bay, photographed by Fred Raynham during the Martinsyde trial flight.

Their aeroplanes, now both tested and ready to go, were evenly matched, and their navigators were equally adept with their instruments and techniques. The respective aerodromes, while each suffered from limitations in size, or ground consistency, or wind direction, had proven themselves good enough for the job of getting the machines off the ground, at least with a light load on board.

Following the Raymor's successful trial, Charles Morgan told reporters that the next flight it made would be the Big Hop. All the teams needed now was a change in the weather, and they would both be ready to get away. It was no longer a contest but a race.

But on April 22, five days after the Martinsyde test flight, the airmen at St. John's received news that changed the parameters of the race. The steamship *Digby*, which had carried the Sopwith team to Newfoundland a month earlier, had returned to its home port of Liverpool and was readying for its next voyage back to the island dominion. It was presently loading its cargo, and the news was that it would be carrying another aeroplane on board: a third entrant for the transatlantic contest. This one was made by the Handley Page company. It was a huge aircraft, a scaled-up version of the twin-engine heavy bomber that Jack Alcock had flown in the Aegean during the war. Its wingspan was 126 feet, almost three times the span of Hawker's Sopwith biplane. The empty weight of the Handley Page was eight tons, over five times that of the diminutive Sopwith; fully loaded with fuel, equipment, and a four-person crew, it would weigh sixteen tons. To get such a leviathan off the ground, it was powered by four Rolls-Royce Eagle engines, compared with the Sopwith's single power plant.

These giant Handley Page aeroplanes had been built for the Royal Air Force to carry out long-distance strategic bombing missions deep into Germany. But the Armistice had been signed before the first Berlin raid could begin, and the aircraft were sitting idle. Frederick Handley Page, like other manufacturers, was looking for civilian orders in a post-war world. Reports of a Handley Page entry into the race had been circulating since earlier in the month, but nobody had known when it would make its appearance, nor who would lead its crew. Now, all was revealed. The aircraft and its team would be sailing for Newfoundland in early May, and its crew leader would be a fifty-five-year-old retired naval admiral by the name of Mark Kerr.

CHAPTER 9

DO BUCK UP AND START

VICE-ADMIRAL MARK EDWARD FREDERIC KERR WAS ABOUT AS close to a British naval establishment figure as you could get. He was born in 1864 at Hampton Court Palace to Admiral Lord Frederic Kerr, later a groom-in-waiting to Queen Victoria, and Emily Sophia Maitland, daughter of a British colonial administrator in Africa and North America. He studied at Stubbington House, near Portsmouth, an exclusive prep school known as "the cradle of the Navy," which counted Robert Falcon Scott among its alumni. After entering the navy as a twelve-year-old cadet in 1877, Kerr rose steadily through its ranks.

By the time of the Great War, Kerr had attained the position of commander-in-chief of the Greek navy. On making the appointment, Winston Churchill, First Lord of the Admiralty, described him as "one of the most gifted and brilliant officers in our service." Then, in 1916, Kerr was given command of the British Adriatic Squadron, based in Taranto in southern Italy. It was in this role that he led a successful coastal bombardment in the Gulf of Trieste that nearly got him shot in the head by an Austrian airman and gave him a lungful of the poison gas phosgene that caused him health problems for years to come.

In the autumn of 1917, Kerr returned to London to help the British government bring its military and naval air wings together to form a unified air force. He had been flying in a personal capacity since 1911. At first, he sat as a passenger alongside Gustav Hamel. Later, he flew with Francis Hewlett, the young man who had been taught to fly at Brooklands by his mother, Hilda. Then Kerr passed his own flying exams, in Greece, in July 1914, three weeks before Britain entered the war. He was the first British naval flag officer to qualify as a pilot.

Throughout the war, Kerr lobbied for aviation to be taken seriously by naval authorities. But, like Lord Northcliffe, he met resistance. He later remembered, "The official and social world considered the aeroplane a toy, aviators reckless madmen, and any married man who left the ground a sort of criminal."

Kerr's work in 1917, helping to found what would become the Royal Air Force, was rewarded the following January when he was appointed Deputy Chief of the Air Staff under Hugh Trenchard. But Kerr could be outspoken and was known for his capacity to make enemies. He locked horns too often with Trenchard over matters of air strategy and found himself sidelined. He no longer fit in the service he'd helped to create, and he stepped down from the job once the Royal Air Force came into being, three months later. In April, he took an air force position in Salisbury, commanding the south-western area. It was hardly a role befitting a man of his status and experience. In October, at the age of fifty-four, he retired.

The idea of being obsolete, like the Berlin bombers now standing idle in the Handley Page factory, didn't sit well with him. He first thought of competing in the transatlantic contest back in 1913, when it was originally announced. The outbreak of war put paid to those plans, but he never lost the ambition. His unanticipated freedom after the war was over gave him the opportunity to try once more.

In November 1918, Kerr approached Frederick Handley Page with the proposition that he would lead an entry into the competition. "My principal qualification for the operation," he later wrote, "was the fact that I was the oldest flying pilot in the world, and so was a good advertisement for any machine." Page agreed to Kerr's proposal. The idea fit with his post-war strategy of building large, reliable aircraft suitable for long-distance civilian air travel. Not for him were the short hops that other firms were planning. He remarked, "We are not interested in short flights. We don't want to take the City man to his home in Balham, but we do want to take you to India and Australia and across the Atlantic."

IT DIDN'T TAKE THE HANDLEY PAGE DIRECTORS LONG TO BUILD Kerr's flight crew. Second-in-command as his co-pilot would be twenty-four-year-old Herbert Brackley. On enlisting in the Royal Naval Air Service in 1915, Brackley received his initial flight training from Jack Alcock at Eastchurch. For the rest of the war, Brackley fought hard and well, coming to excel as a bomber pilot. After the hostilities ended and Brackley found himself looking for work, Kerr quickly secured his involvement in the Handley Page transatlantic mission. He said he felt Brackley had "no superior as a big-machine pilot."

Joining Kerr and Brackley would be twenty-eight-year-old Frederick Wyatt as the aircraft's wireless officer. During the war, Wyatt spent time as the radio man on merchant marine convoys. He also worked with Britain's intelligence service, helping to develop wireless direction-finding, a promising new technique for locating enemy vessels, aircraft, and troop formations from their wireless transmissions. This experience, as well as his hard-won familiarity with the Atlantic shipping routes, made him a sound addition to the Handley Page crew.

The fourth to join the mission would be its navigator, the thirty-one-year-old Norwegian explorer Tryggve Gran. On July 30, 1914, five days before the outbreak of war, Gran had become the first person to fly across the North Sea, from Britain to Norway. It was the longest flight out of sight of land that had ever been made.

Though the crew was now in place, Handley Page was still waiting for government approval to use one of the unwanted Royal Air Force bombers. While negotiations continued, Page and Kerr turned to the most urgent element of the project, as well as the least predictable. Like Sopwith and Martinsyde, Handley Page needed an aerodrome site on Newfoundland. But its requirements were more difficult to meet than those of the other two contestants. The Handley Page bomber, being a lot bigger and heavier than Hawker and Raynham's machines, needed a much longer taxiing run to take off successfully—at least half a mile. Moreover, the disassembled aircraft would occupy countless oversized crates. The route from St. John's Harbour to the aerodrome site would need to accommodate an extensive, heavy, and fragile load.

But here, the Handley Page company had an advantage over Sopwith and Martinsyde. Where the two smaller firms despatched people from their own staff as advance parties—Charles Morgan for Martinsyde and Montague Fenn for Sopwith—Handley Page was able to recruit an agent at St. John's, somebody who knew the territory intimately and could search for the best location to meet the firm's exacting requirements.

Robin Reid was the nineteen-year-old son of Robert and Edythe Reid, leading members of an industrial dynasty which, by the early years of the twentieth century, ran Newfoundland's rail network, mines, pulp plants, a sawmill, a dry dock, a fleet of eight coastal steamships, the electric streetcar system, and the electricity supply network of St. John's. The family company's landholdings covered 2.5 million acres, four times the size of Luxembourg. The firm had almost as many workers on its payroll as did the Newfoundland government.

In 1917 Robin Reid enlisted in the Royal Flying Corps and moved to England for training. He became an exemplary airman. Gaining the rank of lieutenant, he flew machines made by the American Curtiss company as well as Avros and Sopwiths. He spent the final months of the war flying in Italy, and he was there when the Armistice was signed. His service earned him the Italian *Croce di Guerra*.

After his demobilization, Reid was in London. It was early in the new year of 1919 when he was introduced to Mark Kerr by Edward Morris, a member of Handley Page's board of directors. Morris had also, until recently, been the prime minister of Newfoundland, and knew the Reid family well. Kerr asked Reid for his help in finding an aerodrome site suitable for their gigantic machine, and it was a task the young Newfoundlander took on with relish. On February 7, he arrived at St. John's and set to work on his assignment. Driving one of his father's Rolls-

Royce motor cars, he spent the next month ranging over hundreds of miles in search of the perfect spot. The snow and cold weather made his job particularly challenging. "At one time we considered the possibility of using a frozen lake as an aerodrome," Kerr remembered; "but as time went on and we could get no machine, the ice began to melt, and put that suggestion out of the question."

On March 12, a month after he arrived home from the war and two days after his twentieth birthday, Reid left the Rolls-Royce at home in St. John's and headed for the railway station. His train made a seventy-mile journey, first passing south-west out of the city, then rounding the base of Conception Bay, and finally progressing up the opposite side. Reid knew Newfoundland's Avalon Peninsula well. His family had a deep understanding of the region—its topography, its geology, and its meteorology. Just as usefully, the Reids had a vast network of contacts there: people who knew the family of old and who would render any service to assist. With this insider knowledge and the resources of his family to support him, the young airman discovered that the best land for an aerodrome would not be found in the high ground of Mount Pearl, with its unpredictable, gusty winds and its badly drained earth. Nor would it exist in the cramped, rutted ground in Pleasantville, where obstructions hemmed in even the small Martinsyde aircraft. Reid would have understood why the rival firms selected sites close to St. John's, where the ships bringing their crated aeroplanes would expect to dock. They had to limit their search area, and wanted to minimize the distance their bulky crates would have to be hauled. But Reid had the upper hand. He struck lucky in the fishing town of Harbour Grace, twenty-five miles as the crow flies from St. John's.

One reporter, praising Reid's work, dismissed the Sopwith and Martinsyde arrangements as "sketchy and primitive." But this was unfair. Reid had something that neither Charles Morgan nor Montague Fenn possessed. It wasn't just his local knowledge, nor his privileged access to the island's rail network. It was cash. Between the well-connected Handley Page directors and his own family fortunes, Robin Reid was able to draw on deep pockets.

What Reid found held out the promise of being the best aerodrome in Newfoundland. It was a large waterfront plot of land in the east of Harbour Grace, about half a mile in length and 200 yards wide. It occupied a site owned by a New York banker, stretching from a spot close to the town's Gothic-style cathedral, with its impressive twin spires, to Bear's Cove, facing the Atlantic. Entry to the site was only a quarter of a mile from Harbour Grace railway station—part of the network that Reid's family operated—and a dirt road connected the two locations. But there was a downside: it wasn't one single plot. Instead, it comprised a series of farms, gardens, and other properties falling under the leaseholds of at least eight different parties. Heavy stone walls and wire fences divided the site. Mature trees were dotted here and there. There were hummocks and hollows and a couple of ditches. An old barracks building occupied part of the site, long disused but large and inconvenient.

And there were three houses. They were modest homes, but beloved. They were occupied by families who had cared for them for lifetimes; people who had grown up in them and made memories there, memories of when the town's fortunes were in better shape. But this was to prove no obstacle. As one reporter said, "The population is so eager to have the place the scene of a transatlantic start that they have literally offered to move the whole village if it is in the way."

After a week negotiating terms with the site's multiple leaseholders, Reid secured a one-year agreement on the land. He then recruited a gang of a hundred labourers to convert the unpromising plot into an aerodrome fit for the Handley Page. Reid's team started work at Harbour Grace on March 31, the same day that Charles Lester began hauling the crated Sopwith from St. John's railway station to Mount Pearl, and the same day the crated Martinsyde was halfway through its journey from Woking to England's Liverpool docks. The work required at Harbour Grace, back-breaking toil, would take a month to complete. The walls were knocked down. Hundreds of tons of rock were carted away. The large old trees were felled, and their heavy, gnarled roots dug up. The barracks building, no longer wanted, was demolished. Dips and hollows in the ground were packed with rubble, as were the ditches crossing the

One of the houses at Harbour Grace being jacked up and slid away.

site. Then the entire plot was slowly flattened with a weighted roller so heavy, a team of three horses was required to haul it across the ground.

The three houses were each jacked up onto rigid frameworks before being winched away on rails laid on telegraph-pole sleepers to new sites nearby. A reporter who watched the houses being taken away observed, "It was like tearing up the roots of old days and old associations. Despite this, however, their owners were willing to have them removed for the sake of the giant aeroplane that was to re-establish, probably, the town's former prestige."

The resulting aerodrome came to be known among locals as Handley-Page-on-the-Sea. It was a triumph of careful planning, good-will, and ample funding. Just this preparatory work alone cost Handley Page $25,000.

AS THE LABOURERS AT HARBOUR GRACE SPENT APRIL TOILING TO create the Handley Page aerodrome, back in St. John's, the waiting game continued. The weather would not seem to give the flyers a break. The real unknown was the weather patterns over the Atlantic. Daily, the aviators scrutinized Lawrance Clements's latest reports of the transatlantic route; each time, their conversation with Clements frustrated the airmen. The storm belt over the Atlantic seemed stuck there, bedded down for as long as the meteorologist could foresee. There were thick, white fogs that rolled over St. John's from miles out in the ocean. Winds were starting to blow more often from the east, the worst direction, and their patterns remained unpredictable. Rain continued to fall, and one morning, a heavy sleet storm added to the teams' collective misery.

Each new day brought frustration. Morgan later recalled, "It was a big strain—the days of waiting, thinking what the weather would be, the disappointment and strain of wondering what was in store for us all." Hawker and Raynham realized that something would have to give. Each team's constant second-guessing of the weather and the other's intentions was wearing the airmen down. Both feared the consequences of what was described as a "hurried, unconsidered departure." So, they reached an agreement: neither would attempt to leave until both agreed it would be safe to do so.

The decision to collaborate as much as to compete released the pressure on both teams. With preparatory work at the aerodromes now largely complete, the airmen could begin to relax into their enforced stay on the island. But anxiety was soon replaced by a frustrating ennui. Sensing a shift in mood from industry to monotony, one reporter wrote, "Everyone is now living from day to day and has given up trying to prophesy the probable dates of starts. Now that practically no work remains to be done on the machines the airmen find time hanging very heavily, and how to pass the hours is a matter of some difficulty." To break the tedium, Hawker, Grieve, and Fenn would go for long drives from Mount Pearl in a rented Ford motor car. There was much to explore in the striking countryside and coastline around St. John's. There were picturesque fishing villages set in coves and bays. Acres of farmland offered charming vistas. Lakes and ponds dotted the area and snow-topped wooded hills in the distance rose from green valleys. Sometimes, Grieve joined Raynham in the occasional round of golf. The venerable Bally Haly course, located on high ground across the Virginia River

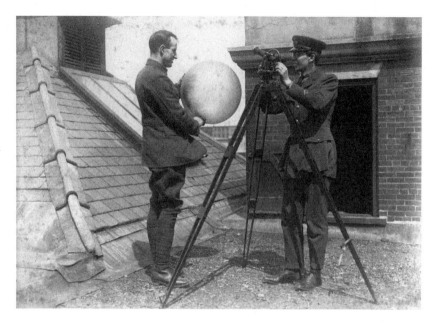

Lawrance Clements (right) testing the wind with a colleague.

from Raynham's Pleasantville aerodrome, furnished views across the Atlantic as far as the eye could see.

Time was really dragging now. "For sheer agony waiting on a job like ours was about the limit," Hawker observed. "Newfoundland is not a fashionable watering-place with a thousand and one things going on."

MOST NEWFOUNDLANDERS—THERE WERE ABOUT 250,000 OF THEM at the time—made their living from the island's natural resources, whether directly or indirectly. In the coastal waters, cod was the most abundant fish, but there were also plentiful stocks of herring, salmon, and lobsters, as well as seals, hunted for their skins and oil, and giant squid with tentacles that could reach thirty feet long. In the interior there were great woodlands of spruce and fir ideal for newsprint, along with forests of oak, elm, ash, pine, and birch. Farmers working the fertile river valleys planted cereal crops, hay, potatoes, roots, and cabbages, while those on the high plateaus to the west of the island raised cattle and sheep. Copper

ore won from mines around Notre Dame Bay generated handsome prof-
its for mine owners. Slate, as good as the best from Wales, was a popular export product, and the island possessed substantial seams of coal
and asbestos as well. There was gold and silver to be found, and in the
1890s vast deposits of iron ore were discovered under Bell Island, twelve
miles west of St. John's in Conception Bay. Before long the Bell Island
mines, exporting their ore to Nova Scotia and Germany for smelting, had
become the most productive in the world. Bell Island ore was so abundant
that one expert said it could be "mined as cheap as dirt."

But an economy based on the sea, the land, and the earth meant Newfoundlanders had to work hard, often in multiple jobs according to the
season. Most lived on low wages. They might have craved recreation, but
they had little spare time or money to enjoy much of it. As such, St. John's
offered unpretentious entertainment. If the fare at the Cochrane began to
pall, the airmen could dine out, but there wasn't much choice. The town
boasted just eight restaurants and cafés, most of them clustered on Water
Street. There was a bowling alley there as well, which might have helped
pass an hour or two. There was no liquor to be had on the island, at least
not legally, as the dominion was still operating under wartime prohibition laws. This didn't bother Hawker, who was a lifelong teetotaller, but it
meant the others—described, by Hawker, as "the unfortunates who still
found total prohibition not quite to their taste"—had few options but to
join him for ices at Harry Peddigrew's drugstore at the corner of Prescott
Street and Military Road.

On at least one occasion, Hawker and Grieve attended a dance.
The Blue Puttee dance hall, three blocks from the airmen's digs at the
Cochrane, opened its doors for informal afternoon and evening dancing
earlier that year. Its advertisements in the local press boasted "an up-to-
date Victrola and the latest dance records," and advised that "no lady
may feel the least hesitation of attending any session."

But the biggest nights out for the airmen and ground crews were their
visits to the town's half-dozen movie theatres, which included the Nickel,
the Crescent, and the Star. The latest attraction, which had its grand
opening only the previous month, was the Majestic. It was a large, new

building on the corner of Duckworth Street and Theatre Hill, a few steps from the electric streetcar line. Inside was an auditorium and gallery which, combined, seated 1,100 moviegoers. They sat in spacious tip-up seats, with plenty of legroom, arranged carefully among aisles so that latecomers wouldn't block the presentation as they found their places. The latest imported movies, projected by the most modern equipment—a Power's machine driving a flicker-free Metz lamp—were shown on a bright, silver-coloured screen measuring thirteen feet by eleven, flanked by towering fluted columns topped with attractive capitals. The panelled and coved ceiling dampened the acoustics. Electric lighting, good ventilation, and steam heating guaranteed patrons' comfort. On opening night, with the dominion's governor and the Catholic Cadet Corps band present, two thousand eager St. John's residents mobbed the building's entrance hoping for a seat inside. By the time the airmen were there, the Majestic had become one of the town's most popular entertainment spots. It was, in the words of one approving reporter, "the most up-to-date and healthy public place of amusement that the city has."

By now, the absence of any discernible progress at the aerodromes was starting to frustrate those watching from afar, as well as those on the teams themselves. Increasing quantities of mail had begun to arrive at St. John's from members of the public on both sides of the Atlantic, tendering good wishes. There were letters, poems, autograph requests, offers of assistance, and weather forecasts from those fancying themselves as amateur meteorologists. Sometimes the good wishes were wrapped, for fun, in sterner words. One letter, sent to Mac Grieve by two British cable operators working in New York, said, "Do buck up and start—we cannot stand the suspense much longer." When he read the message, Grieve was said to have retorted that the women's suspense was nothing compared to his own.

The mailbag also brought occasional gifts from well-wishers: mascots and charms to bring the airmen good luck. One present that Raynham received was a wooden parrot, named Emma, which he attached to the cockpit of the Raymor. Grieve planned to carry a woman's handkerchief and a posy of white heather he had been sent. Hawker, asked if he would

The Majestic Theatre, St. John's, about 1920.

be taking a lucky charm with him on the flight, replied that Grieve was
the only mascot he believed in.

IT HAD BEEN FRUSTRATING FOR MARK KERR TO WATCH THE
Sopwith and Martinsyde teams prepare their aeroplanes and ship them
to Newfoundland, all while Handley Page waited for a government decision on whether it could use an aircraft sitting idle in the works. By late
March, though, consent had been given, and the chosen machine was
quickly converted for the transatlantic attempt.

The Handley Page and its eighteen-strong team travelling on *Digby*
reached St. John's on Saturday, May 10 after a seven-day voyage. The
crates were offloaded, with some difficulty, onto a long train of Reid company railcars waiting on the dockside. First to be winched off the ship
were the six enormous, iron-strapped wooden cases containing the aircraft's wings, fuselage, engines, and propellers, which had been stowed
on the open deck. Each case weighed between two and a half and five
tons. Then, the dock workers turned their attention to the smaller crates
stowed in the ship's hold. There were over a hundred of them. It took the

rest of the weekend for the shipment to be extracted, lowered onto the railcars, lashed down, and made secure. On Monday, the train began to make its slow way along the same seventy-mile route to Harbour Grace that Robin Reid had followed two months earlier.

It was a challenging journey to complete. At one point, the locomotive was approaching a bridge that crossed over the tracks. Normally, there would be ample clearance for a freight train to pass underneath. On this occasion, the driver realized that the crates carrying the aircraft's wings were too tall to clear the bridge and were about to collide. After bringing the train to a halt, he gathered the men travelling with the shipment. Together, they grappled the oversized crates off their railcars, which was not an easy task as they had no lifting equipment. The driver took the train through to the other side and pulled up once more. Then, each of the huge wing cases was slid directly along the steel rails. They just cleared the bridge. Getting them back onto their railcars was a feat of brute force as much as careful planning.

Matters hardly looked up once the train pulled into the station at Harbour Grace. With nothing but heavy timber railway ties to assist the process, the crates were laboriously slid off the railcars before being stacked up ready for the quarter-mile journey by dirt road to the aerodrome. This task appeared, in the words of one local reporter, "almost herculean." But a novel idea came to the men. They unpacked the crates containing the aircraft's heavy undercarriage, with its four giant wheels, each over four feet in diameter. Then, they lashed the axles together with long telegraph poles, forming a makeshift dolly that allowed them to wheel the remaining crates down the road and across the grass of the aerodrome with little difficulty.

It took the team until May 16 before all the crates had been moved safely. Then the task of assembling the enormous aircraft began. There was no hangar remotely large enough to accommodate such a beast under cover, so assembly took place in the open. It was a slow, painstaking process, carried out most often in foul weather. The engineers had to bunk down each night in a railcar parked up at the station, or in rooms in Harbour Grace town.

Dollying a crate using the Handley Page's undercarriage wheels.

Vice-Admiral Kerr left them to it. He was staying at the grand St. John's residence of Robin Reid and his parents, and the Reids had put their fleet of Rolls-Royce motor cars at Kerr's disposal. They ferried him to the meteorological office in downtown St. John's each day; he also went to see the Sopwith and Martinsyde aerodromes in Mount Pearl and Pleasantville, sites he described as "bad bits of ground." Later, he claimed to recall trying to talk Hawker and Raynham into abandoning their fields and joining him at Harbour Grace. However, "difficulties arose which could not be got over," he wrote, after the event. So, Handley Page remained aloof out in Harbour Grace, and Sopwith and Martinsyde stayed where they were.

The pressure on all three teams ratcheted right up, however, when the Vickers party arrived in town.

CHAPTER 10

HANG THE WEATHER!

JACK ALCOCK WOULD LATER SAY HE WAS "IN THE GAME FROM THE very jump." As soon as he returned to England from the war, he began to look for an aircraft he could fly across the Atlantic. "It was not that he wanted to advertise himself, or to become a popular hero," wrote his friend, the aviation journalist Charles Grey, "for he is, and always has been, one of the quietly confident sort who does much and talks little."

Early in the new year of 1919, Alcock attended a meeting of the Royal Aero Club, where he ran into an old acquaintance, Warren Merriam. He and Merriam first met as flyers at Brooklands before the war and quickly became good friends. They encountered each other again at Eastchurch once hostilities had broken out: both were instructors at the flying school there. But Alcock's posting to the Aegean in 1916 meant the two men didn't see each other for more than two years, and they had a lot to catch up on when they were finally reunited after the war's end. Their conversation soon came around to the *Daily Mail* prize. A rumour was spreading among the flying community that the Vickers company wanted to enter the competition but had not yet found an aviator. Merriam told Alcock that he wanted to approach Vickers and take a shot at the prize; rumour had it he would be in with a good chance. Then Alcock replied that he wanted to take part in the competition as well.

Brooklands flyers always competed, sometimes fiercely, but they looked out for each other as well. Old friendships forged in those pre-war days of aviation were resilient. The two men decided to let fate choose which of them would approach Vickers, and they tossed a coin. Jack won.

THE VICKERS ARMAMENTS AND ENGINEERING COMPANY DIVERSI-fied into the aeroplane game in 1911, when the military potential of heavier-than-air flight was becoming evident to all but the most reaction-ary authorities. That year, Vickers began to design and build aeroplanes at its factory in Erith, Kent, sometimes test-flying them at Brooklands. In 1912, the company founded a flying school at Brooklands, taking sheds adjacent to Hilda Hewlett and Gustav Blondeau. The Vickers school, which used aircraft manufactured by Hewlett's firm, quickly became prolific. In August 1914, much of the firm's aircraft production was transferred from Erith to its plant in Crayford, about two miles south. Six months later, the firm decided it needed a second factory, and selected the site of a former motor car works just outside the main Brooklands racetrack. The job of setting it up was given to a man named Percy Maxwell-Muller, who became the works manager; Archibald Knight, known as Archie, was his assistant. Before long, the Brooklands factory opened, and in June 1915 it began turning out B.E.2c aircraft ready for despatch to the fronts.

A new opportunity for the company arose in July 1917. Daytime air raids over London the previous month had spurred the British govern-ment into planning the long-range strategic bombing of Germany. To carry it out, heavy bomber aircraft would be needed in large numbers, and an order for prototypes was made with Vickers. Its chief designer set to work on blueprints, and just four months later the prototype was ready. It performed superbly in tests, and big production orders were placed. Before long, the Vimys, as they were named, began to roll out. But the war ended before they could be put to their intended use. Like the big Handley Page bombers, they were suddenly surplus to requirements.

Some of the completed aeroplanes were sold to the peacetime Royal Air Force, and others found buyers overseas. But, as Tom Sopwith had found, the bottom had dropped out of the market for military aircraft. For Sopwith, the post-war prospects of staying in business as an aeroplane manufacturer, at least in the short term, looked bleak. But Vickers, with its long-range heavy aircraft, was, like Handley Page, pinning its hopes on a new market. The firm's solution was to turn the wartime Vimy into a peacetime product—a passenger-carrying airliner. It would come to be known as the Vimy Commercial. But there was no market yet for such products, and nobody then knew whether there ever would be— in 1918 the public didn't even know what an airliner was. So, the *Daily Mail*'s transatlantic contest was a potential lifesaver because it would help build the idea of civilian air travel in the public's mind. No wonder Vickers was keen to enter a Vimy into the competition.

To stand any chance of success, it would need the best pilot at the controls. Then, into the Vickers factory at Brooklands walked Jack Alcock, fresh from the Aero Club meeting where he had discussed his future with Warren Merriam. He was still on leave, and he wanted to catch up with his old friends, Percy and Archie.

The three airmen had grown up together at Brooklands before the war separated them. Percy Maxwell-Muller took his flying lessons at Maurice Ducrocq's school, where Alcock worked. In 1913, within weeks of qualifying, he had built his own aeroplane, which Alcock test-flew for him. Archie Knight had taken his flying tests the month after Alcock arrived. In November 1912, Alcock and Knight competed in a two-lap speed race around the Brooklands circuit. It was Alcock's first competition, and he won, beating Knight into third place. It was an early indication to Knight, and to the crowds watching that day, that Alcock was a talent to watch. A lot had happened since then.

As the three friends caught up at the Vickers works that day in early 1919, Maxwell-Muller casually asked Alcock, "How would you like to fly the Atlantic?"

Alcock reportedly replied, "Ripping. I am certainly keen on that if you can get the machine ready."

Maxwell-Muller took Alcock into the factory and they approached the Vimys waiting on the shop floor. Alcock pointed to one of them.

"There's the machine that can be converted," he told Maxwell-Muller. The job was his. But they would also need a navigator.

BEFORE THE WAR INTERVENED IN THE HAPPY PROGRESS OF HIS life, the Manchester engineer Ted Brown had become an adept amateur in navigating the Atlantic. On his frequent transatlantic steamship voyages on business for British Westinghouse or to visit his American family, he never wasted an opportunity to advance his technical knowledge. Having developed an interest in aerial surveying and navigation during his time with the aviators of South Africa, he would invariably ask the ships' captains for permission to visit them on the bridge, where he would discuss astronomical techniques with the navigating officers and make sextant sightings of the sun and the stars.

Having been born in Glasgow to American parents and brought up in Manchester and Pittsburgh, Brown felt no fixed sense of home. "I caught neither a Scotch nor an American nor even a Lancashire accent," he later said. It was as if he lived *on* the Atlantic: straddling it, with one foot in North America and the other in Britain. He certainly knew the ocean well.

After the Armistice, when he was at home in Manchester on sick leave, Brown started anticipating his formal release from the Air Ministry, to which he was nominally attached. His convalescence bored him. He stayed in the house most days, working on navigation articles for engineering journals. Once he recovered, he and Kathleen wanted to move to New York, where he hoped to find an engineering job. But the idea of navigating a transatlantic flight before they settled down had become lodged in his mind as the war concluded, and he couldn't shake it out. He started approaching aeroplane manufacturers, offering his services as a transatlantic navigator, but none would take him on. He dropped the idea for a while and began, like all other airmen leaving wartime service, looking for a permanent job. He even tried to sell one of his clever inventions—a new design of fuel pump—to Tom Sopwith and Harry Hawker. But they turned him down. Brown found it hard not to take such rejection personally.

One day, he called on the Vickers factory at Brooklands for an interview with Percy Maxwell-Muller. He had been given an introduction by David Kennedy, Kathleen's father and his former boss at the Ministry of Munitions, who had found out there was an engineering job vacant.

Brown later recounted what happened. "While I was talking with the superintendent, Captain Alcock walked into the office. We were introduced, and in the course of conversation the competition was mentioned. I then learned, for the first time, that Messrs. Vickers were considering an entry, although not courting publicity until they should have attempted it. I sat up and began to take notice, and ventured to put forward my views on the navigation of aircraft for long flights over the sea. These were received favorably, and the outcome of the fortunate meeting was that Messrs. Vickers retained me to act as aerial navigator."

The Air Ministry agreed to Brown's reassignment to Vickers, and he lost no time in getting started on the project. With the ministry's help, he began acquiring the navigational instruments and other specialist equipment the aircraft would need. He also took time to get to know Jack Alcock better and made sure the pilot was consulted over every arrangement. In the meantime, the rest of the Vickers workforce made resolute progress on converting the Vimy from a bomber to a transatlantic machine.

The period leading up to the Vimy's test flight was one of intense and pressured activity, as each specialist team put the finishing touches to their part of the machine. One group made the structural fabric used to skin the aircraft fuselage, wings, and moveable control surfaces. Leading this team was thirty-year-old Annie Boultwood, who originally trained as a bookbinder before turning to aviation. At Vickers, Boultwood ran a department of three hundred women. Much of their work involved sewing the heavy canvas fabric as it contorted around the aircraft's curves, supporting struts, and the complex lattice of bracing wires and control cables. The women also applied the cellulose-based lacquer known as dope, which stiffened the fabric and drew it tightly onto the supporting framework, and shellac, an insect-derived protective resin. Years later, Boultwood recalled the eighteen-hour working days leading up to the test. She also remembered the striking appearance of the aircraft when it was finally ready to fly. "There was shellac on the ribs, on the wings and on the tailplane elevator," she said. "All the dope was silver."

The test flight took place at Brooklands on Friday, April 18, the day after Fred Raynham, 2,300 miles away, had taken the Martinsyde up for its own trial over St. John's. Boultwood watched the scene as it unfolded at the Surrey airfield that day. "I saw it take off for that first flight," she recalled; "it looked beautiful, like a bird." The flight, with Alcock, Brown, and a heavy load of fuel on board, was a success.

After final tests and adjustments had been completed, the Vimy was disassembled for the journey to Newfoundland. Along with twenty-three drums of fuel, the aircraft was sent with a party of mechanics and riggers to the London Docks, where it was to join the steamship *Glendevon*, bound for St. John's. It was due to sail on May 3, but it got away late owing to a strike at the docks, not leaving until May 13.

The rest of the party went on ahead. On May 2, Ted Brown kissed Kathleen goodbye at Woking station as he headed to Southampton. There, he, Jack Alcock, and four Vickers engineers and specialists boarded the grand Cunard liner *Mauretania*. The vessel had not yet been released from wartime government charter. It was being used to repatriate American and Canadian troops heading home from the war—twenty thousand of them since the Armistice had been signed. A strike among boiler workers at the Southampton docks delayed departures there as well, making it a frustrating time for the Vickers party, who just wanted to get on with the job. The vessel finally left Southampton on May 4.

Once on *Mauretania*, the Vickers men got acquainted with their luxurious temporary home. It was five years since Brown had been on a transatlantic liner. True to the habits he had developed before war broke out, he approached the ship's commander, forty-nine-year-old Arthur Rostron, and asked to visit the bridge to practice his navigational techniques. As captain of the Cunard liner *Carpathia* in 1912, Rostron had raced his ship through the treacherous ice floes of the North Atlantic to aid survivors of the *Titanic* disaster, rescuing more than seven hundred people. In 1915 he saw extensive action commanding troop and hospital ships in the Battle of Gallipoli. His command of *Mauretania* in its guise as a transport ship later in the war further added to his experience, making him one of the most capable mariners on the seas. Such knowledge was valuable to the aspiring aerial navigator Ted Brown, and he was glad when Rostron assented to his request for the freedom of the bridge. The captain also gave his younger and greener passenger "much good advice," as Brown was later to recall. For his part, Rostron remembered Brown and Alcock as "good fellows."

The passage across the Atlantic took five days. *Mauretania* wasn't putting in at St. John's; it was heading for Halifax, Nova Scotia. The Vickers men would have to make their own way to St. John's from there. They arrived at Halifax in the early morning of May 9, and it would take almost as long again to reach St. John's. First there was a long overnight rail journey from Halifax to North Sydney, a port town on Nova Scotia's north-east coast. Then they caught a steamer that took them over the water to Port aux Basques, on the south-western tip of Newfoundland. Finally, there was a day-and-a-half rail journey on a wide, sweeping arc of the island to the terminus at St. John's. They finally reached there early on May 13.

With their aeroplane and stores just starting out across the Atlantic on the *Glendevon*, the most important task for the advance party was to find suitable land for an aerodrome. The original plan had been for Alcock to head for Newfoundland in February and carry out the scouting job for Vickers that Charles Morgan, Montague Fenn, and Robin Reid had done for their own respective companies. But things had not worked out as planned. For some reason, possibly to do with his demobilization, Alcock couldn't go out that early. Now he would suffer the consequences. The Vimy aeroplane, being heavier than the Sopwith and Martinsyde entries, would need a longer runway to get away safely— about 500 yards. This was nothing like the 900 yards the lumbering Handley Page required, but it was a problem, nonetheless. Alcock and Brown would have to get the lay of the land quickly. Wasting no time, the

pair hired a motor car and drove six miles west out of the town centre to
Mount Pearl, where the Sopwith aerodrome was sited. They wanted to
see if there was any suitable ground nearby.

It was a dispiriting drive. As they searched for a field or clearing, they
discovered that level and even ground was alien to the area. They saw
rolling hillocks. They saw soft, thin soil with poor drainage that turned
boggy in the rain and sleet, and when the snowfall melted. They saw
stones and boulders strewn across every field, each lying ready to trip up
a taxiing aeroplane and tip it sideways, or forward onto its nose, without
warning. And they saw stone walls and fences that divided the land into
small plots. Nothing was suitable.

Returning to St. John's that evening, the pair joined the rest of their
team at the Cochrane Hotel, where they would be staying. They met the
other residents: Harry Hawker, Fred Raynham, Mac Grieve, Charles
Morgan, and the newspaper correspondents who had been covering the
story since March. For Alcock, Raynham, and Hawker, it was a happy
reunion. The new arrivals caught up on the latest developments. No doubt
they discussed the frustrations of bad weather, which was still preventing
a departure. The arrival of the Handley Page crew three days earlier, a
group now starting their assembly at Harbour Grace, would have been
fresh in the Cochrane residents' minds. After the hurry among Hawker
and Raynham to get their aeroplanes ready and away, the weather had
levelled the field. Now it seemed just possible that Mark Kerr could win
the contest. But the most pressing questions Alcock and Brown wanted
to ask their friends concerned ground conditions. The answer they got
back from the men was clear, as Brown remembered: "that the only suitable patches of ground had been appropriated, and that we should find
no others near St. John's."

Alcock and Brown spent the next week searching for a spot. They gave
up hiring cars each day and bought a second-hand seven-seat Buick in
bad shape. They beat it up some more. For hundreds of miles, they drove
the rutted dirt roads up and down the Avalon Peninsula. Each time they
saw a patch of land that looked as if it might suit their purposes, they
would get out of the vehicle and pace out the length and width of the

meadow or field. Then, invariably, they would return to the car disappointed to have found that the clearing was too short or faced the wrong way for the wind. In the evenings, they joined the other airmen in visits to the movie theatres or played cards at the Cochrane. Sometimes they just strolled around St. John's, exploring the town.

While the Vickers team searched for an aerodrome site and waited for the delayed arrival of their aircraft, and while the Handley Page crew continued the assembly of their giant bomber at Harbour Grace, the Sopwith and Martinsyde groups at Mount Pearl and Pleasantville greeted each new day with hope for good weather and a chance to get away. There were times when the local conditions were good, but reports from mid-Atlantic told of storms. At other times, the ocean conditions were reportedly acceptable, but wind and fog over St. John's prevented a safe take-off. Wednesday, May 14 looked promising. A full moon that night would have lit the way well. But it wasn't to be. "Mr. Harry Hawker's machine is practically ready for the 'big hop,'" reported the *Daily Telegraph*, "but adverse weather and ground conditions have prevented him making a start." Hawker and Raynham felt like they were being held "like greyhounds on the leash," as one reporter put it. They'd come this far. They needed the weather to be just right.

That all changed on Friday the 16th. News reached the airmen that three American navy seaplanes were waiting sixty miles away at Trepassey, on Newfoundland's southern coast. They were preparing to take off for the Azores, and then for Europe.

KNOWLEDGE OF THE AMERICAN SEAPLANES HAD BEEN NAGGING at Hawker and Raynham for weeks. The aircraft were known as "NC's," short for Navy-Curtiss, though most took to calling them "Nancies." Four had been planned in a joint venture between the navy and the Curtiss aircraft company for use in the war, but the first had been completed just too late to see action. After the war ended, with construction on the remaining three machines still ongoing, a decision was made that the fleet, once complete, would attempt the Atlantic.

It was nothing to do with the *Daily Mail* contest. The Nancies would make the crossing in a series of hops, rather than the single flight that would win Lord Northcliffe's cash. As seaplanes, they were designed to land on and take off from water, which meant they could come down in the Atlantic if storms in the air got in their way, or if they developed engine trouble or mechanical problems. Moreover, by making the crossing in a series of shorter flights, they could refuel, take on provisions, and be attended by maintenance engineers at stops along the way. In a further line of support, the navy was mobilizing a fleet of forty-five vessels to line the route, anchored in the ocean at fifty-mile intervals. If any of the seaplanes had to come down in a hurry, help would never be far away.

By the time the seaplanes were ready, the fleet had shrunk from four to three. A storm in March caused extensive damage to the wings of NC1. Rather than manufacturing new ones, NC2 was turned into a donor aircraft, and its wings replaced the damaged ones on NC1. By the middle of May, all three completed seaplanes—NC1, NC3, and NC4—had arrived safely at Trepassey.

Hawker and Raynham were competing to be the first to fly nonstop across the Atlantic. The American seaplanes were not in that race. Yet there was a matter of honour at stake. In the minds of the public watching on both sides of the Atlantic, small details like refuelling stops and naval backup vessels hardly mattered. This was a wider competition. It was a contest for national prestige, and this was what had nagged at Hawker and Raynham as their own attempts to take off were delayed. They weren't just flying for themselves and for their aircraft companies. They were flying for the honour of Britain.

In newspaper interviews given during the previous few days as work on the aircraft at Trepassey advanced, the British teams had been dismissive of the American enterprise. Hawker told reporters he would wager that "any fast steamer leaving New York on the same day as the 'Nancies' of the navy will beat them to England." Morgan, noting the string of guard-ships that would line the American route, said that the American pilots "might just as well fly the English Channel fifty times,

Navy-Curtiss seaplane NC1 being moored at Trepassey, May 1919.

an ordinary exploit these days. The flight proves nothing practically or theoretically."

The *New York Times* reporter staying at the Cochrane, who had been taunting Hawker for days by accusing him of having cold feet, was unimpressed by this criticism. He wrote, "There is a strong feeling here that Great Britain has been caught napping and is being outstripped by the Americans in the race for precedence in transatlantic air competition." The correspondent for Lord Northcliffe's *Daily Mail* wrote that the American aircraft "may go at any moment," doubtless compounding Hawker and Morgan's frustration.

With the war of words between Britain and America heating up, the Sopwith and Martinsyde airmen had hoped to be up and away before the Nancies even made it to Trepassey. Now that the seaplanes had arrived there, the British teams could only hope they would be delayed by the same weather keeping their own aircraft on the ground. Word was that the American contingent expected a long delay as well. But on the morning of Saturday the 17th, the British airmen woke to the unwelcome news that the American seaplanes had left Trepassey's narrow harbour the previous evening.

Hawker recalled, "All day long we waited for news with, I need hardly say, great anxiety, for Raynham and Morgan and Grieve and I had already made up our minds." That morning, the four men sat down in a quiet corner of the Cochrane to work out what to do next. It did not take them long to come to a judgement on the matter. They calculated that if the Sopwith and Martinsyde aeroplanes took off soon after the Nancies reached the Azores, the British would still have time to beat the Americans to the finish. The blue riband of aerial navigation was still just within their grasp.

Eventually, their anxious wait for news of the Nancies was ended. That evening, an unconfirmed report reached St. John's that the American seaplanes had arrived at the Azores. At the Cochrane, a fateful decision was reached. "After that there was only one thing to do, namely, get as much sleep as we could, for we didn't intend to have any the following night," Hawker noted. "We meant having something more interesting on hand."

THE NEXT DAY, SUNDAY, MAY 18, WHILE RAYNHAM AND MORGAN began making their way to Pleasantville, Hawker and Grieve drove the six miles from the Cochrane up to Mount Pearl. The Sopwith aircraft was brought out of its wooden hangar. It was fuelled up, its oil and water tanks were filled, and its engine was tested one more time. Food and drinks were loaded aboard, along with navigational charts and instruments. Mail for delivery to Britain, including a letter from the governor to the king and one from the dominion's prime minister to his British counterpart, was sealed in a bag and put into the cockpit.

During the day, as word spread that the Sopwith and Martinsyde aviators were planning to depart, crowds began to arrive at the two aerodromes. Thousands headed for Pleasantville to see the Martinsyde team. Being closer to the town centre than Mount Pearl, it was an easy walk for residents. Besides, Charles Morgan had become the town's darling ever since his successful scouting visit back in February. Locals had taken just as well to the youthful Fred Raynham. Nonetheless, about two hundred spectators—those with motor cars or horse-drawn carriages—drove up to Mount Pearl to wish the Sopwith airmen on their way. Among the onlookers there were Mark Kerr and Herbert Brackley, the recently arrived Handley Page pilots.

At noon, all four airmen went to the meteorological office on Duckworth Street to get the latest weather reports. It was more a formality than anything else. Hawker, at least, had little interest now in hearing what Lawrance Clements had to tell them. "Hang the weather! I go this afternoon though it leads me to the Pacific," he was heard to say at the office.

Returning to Mount Pearl, Hawker and Grieve ate lunch with friends before making their final preparations. At 3.05 p.m., as technicians started the aircraft's engine, the airmen pulled on their flying suits. They shook hands with spectators and received fond farewells in return. Then Hawker told onlookers, "I have a perfect machine for the trip and the engine is the best in the world. I am confident we shall get across." Kerr and Brackley bade their goodbyes. Kerr later recalled, "Hawker, as usual, was just as cheerful as if he was going for a joy-ride." Grieve, addressing friends waiting by the aircraft, simply said, "See you in London."

Then the pair climbed into the cockpit. It was time to go. With the Rolls-Royce engine roaring, Hawker shouted, "Tell Raynham I'll greet him at Brooklands." He raised his arm and waved for the chocks to be pulled out from under the wheels of the undercarriage. Then he opened the throttle. The machine jumped forward, starting to bounce and lurch as Hawker steered it diagonally across the L-shaped field to meet the 20-mile-per-hour wind. The crowd at Mount Pearl shouted cheers. As the aircraft picked up speed, it started to roll from side to side with the uneven ground. Hawker kept it moving steadily across the field. The weight of fuel and oil demanded a much longer take-off run than the trial flight the previous month, and Hawker would need every inch of ground available to him.

Mac Grieve and Harry Hawker in front of the Sopwith Atlantic.

As the machine gained speed, the drainage ditch and fence at the edge of the field started to close disturbingly on the oncoming aircraft. Beyond the fence were trees. Still, Hawker held his course, building as much speed as possible. For forty-five seconds, the biplane bounded and skipped across the thin Mount Pearl earth. With inches to spare between the aircraft's wheels and the drainage ditch, and with the cheers of the crowd lost in the din of the engine, the Sopwith hit a bump—and took off. The heavy wheels of the aircraft just skimmed the tops of the trees that sat in a ring around the aerodrome, and then it was away. As the onlookers exhaled sighs of relief, Hawker pushed the aircraft into a steep climb, circling to gain height. Then he throttled down and made a north-easterly course towards the ocean.

The sun shone on the biplane's wings through a sky as clear and bright as any the airmen had experienced. Those St. John's residents who had not gone to one of the two aerodromes that afternoon hurried into the streets of the city as they heard the drone of the Sopwith's engine overhead. Looking skyward, they saw the aircraft's spinning propeller shining in the sun as brightly as a searchlight. They cheered as one. A reporter observed, "Every lip gave expression to the hope that the journey would be made in safety."

At 2,000 feet, the airmen passed over Pleasantville and looked down at the Martinsyde aerodrome. It was now packed with onlookers; some estimates put the figure at ten thousand. In a patch of space at the centre of the great throng, they saw the Martinsyde aircraft. Beside it stood Fred Raynham and Charles Morgan, who raised their arms to the sky and shouted. Hawker and Grieve blew kisses in return.

Soon, the Sopwith was flying over the high, white cliffs that separated St. John's from the sea. As soon as the rocks had been passed, Hawker pulled the trigger that released the aircraft's heavy undercarriage and watched it tumble down to the water. After five more minutes, with the weight and wind resistance of the undercarriage having been lost, the aircraft had risen to 4,000 feet. Before long, it could no longer be seen by those watching through telescopes from Mount Pearl. At a little before 4 p.m., travelling at 100 miles per hour with a tail-wind hastening it along, the aircraft passed out of sight of St. John's, still climbing.

There was nothing more for onlookers to see at the Sopwith aerodrome. The journey from Newfoundland to Ireland was expected to take about twenty hours. The crowds at Mount Pearl began to disperse. Newspaper correspondents went off to file their copy. Reports of the successful take-off were flashed along the news wires on both sides of the Atlantic. Night editors began setting the stories in type.

At the English village of Hook, a message was passed to Muriel Hawker, telling her that Harry had departed safely. She stepped outside their house and began hanging up flags—including the Australian flag, which she mounted over the garden gate—in readiness for his hoped-for arrival. Then she went back inside. As darkness fell, she looked out of the window at the moon. It was on the wane, but it was a comforting sight, nonetheless. As each hour of her husband's expected journey elapsed, she marked off a piece of paper as a record.

The next morning, the Sopwith departure was the splash in all the dailies. The *New York Times* wrote, "All the people of this country will pray that fortune may favor and success crown the brave venture of Harry G. Hawker and his companion, Lieut. Commander Grieve. . . . None but men with hearts of oak would take the risks they take, for all depends upon the unceasing efficiency of their engine and the staunchness of their plane." The *Daily Telegraph* said, "While London slept, Mr. Hawker, with his companion, in an epoch-marking adventure, was driving through the dark Atlantic skies at something like 100 miles per hour. Anxiously they looked for the dawn. Where did it find them? Eagerly will news of their progress be awaited." Back in St. John's, the *Daily Star* wrote, "Hawker and Grieve are among the world's bravest men. They are the type of Britisher who have eliminated the word fear from their vocabulary." Northcliffe's *Daily Mail* said that "The great adventure was at hand."

But some reports expressed concerns about Hawker's decision. The *Manchester Guardian* wrote, "The general feeling to-night is anxiety that Hawker may have allowed himself to be influenced by the success of the Americans reaching the Azores, and begun his immeasurably greater task without waiting for the most favourable conditions."

As the hours of Monday the 19th went by, more and more people around the world heard that the Sopwith team had made it safely out of Newfoundland. The day passed in an air of great excitement. During the afternoon, the evening papers assembled their stories. At St. John's, the *Evening Telegram* wrote, "By the time this appears in type, it is hoped that the landing on the other side will have been made. Given favourable conditions and providing no accident overtakes either men or machine the journey should now have been safely accomplished."

With the press serving up vivid descriptions of the scenes at Mount Pearl, it seemed like every soul on Earth was holding on expectantly for the first reports from Ireland.

They waited . . . and waited. With the world watching, the expected time of Hawker and Grieve's landing in Ireland passed.

They never arrived.

CHAPTER 11

IS RAYNHAM ALRIGHT?

ONCE THE SOPWITH HAD PASSED OVER THE CLIFFS OF ST. JOHN'S and disappeared, all eyes that afternoon turned from Mount Pearl to Pleasantville. The Martinsyde departure was due to take place there at any moment. There were so many spectators crowding the lakeside aerodrome that police officers had to stop the closest ones from grabbing hold of the aeroplane. Not that Fred Raynham and Charles Morgan showed much concern. One reporter observed, "Both were in high spirits and chatted and joked with friends as if they were to take part in some trivial athletic contest rather than in a race with life and death."

After the team had spent so many weeks working on the aeroplane, it was in perfect shape. Now the only thing that mattered was to make a safe departure from the aerodrome, and then—in their slightly faster aircraft—they would be able to catch up with Harry Hawker and Mac Grieve somewhere over the Atlantic. That, at least, was the plan.

Ever since the thirty-year-old Morgan arrived at St. John's on his scouting trip that February, he had taken to the people of the city, and they had taken just as readily to him. He was a tall, sharply dressed man, with an open face, dark hair, and blue-grey eyes that seemed to shine with good humour. In photographs, his half-smile suggested a mischievous character. His war injuries marked him out for respect and admiration, but he was reluctant to accept sympathy. "He was just a schoolboy

Charles Morgan in Newfoundland in 1919, ready to fly the Atlantic.

playing a great game," one reporter said. He joked about his prosthetic leg and told tales of his experiences during the war. If his nightmares and depression continued to dog him during his Newfoundland adventure, he did not let on. Though it was clear to everybody that he was serious about aviation and navigation, he never seemed to take himself too seriously. Unlike Hawker, Morgan welcomed visitors to his aerodrome— particularly journalists, whose questions he answered readily. When locals learned that he planned to start an airmail service between St. John's and Montreal after the contest was over, they were delighted. Even children adored him. One newspaper wrote, "The school boys all knew him and they delighted to touch their caps as he passed and he never failed to respond. The little girls knew him too, and he was their idol."

Raynham was a more reflective soul. In one reporter's words, the twenty-six-year-old was "rather reticent and disinclined to talking about either himself or machine." But he was popular among St. John's residents, nonetheless. He naturally hung back from the crowd and allowed his co-pilot to take the limelight. But everybody could see he was an outstanding airman.

When the Sopwith biplane carrying Hawker and Grieve buzzed overhead and made for the Atlantic that afternoon, Raynham was relieved to see his old friend Harry get away safely. No doubt he recalled the times they had flown together in the past. One local reporter picked up how close Hawker and Raynham were to each other, despite their differences in temperament. He wrote, "A strong bond of affection existed between them and no men held either of them in higher esteem than they did each other. To Raynham Harry Hawker was the best flier in the world. Raynham in Hawker's estimation was unsurpassed as a birdman."

Raynham had not greeted that lunchtime's weather reports with the same equanimity that Hawker had, though. The Pleasantville field was more confined than Mount Pearl, and Raynham had fewer options for the direction in which he could take off. The best direction was into the wind: this created the greatest lift on the wings of the aircraft, and a heavy machine could get away with a shorter taxiing run. But Raynham would have to take off with the wind coming from the side that afternoon, generating much less lift. It would need to be a long run.

The Raymor had already been pulled out of the canvas hangar and hauled into position at the end of the field. Chocks were placed under the wheels, with long lengths of rope attached to allow them to be pulled away from the safety of the sidelines. Morgan helped Martinsyde officials marshal the crowds away from the line of the take-off run, though the curiosity of the spectators made this a challenging job. Food and hot drinks were placed aboard the aeroplane, and Emma, Raynham's wooden parrot mascot, was fixed carefully inside the cockpit. At 3.50 p.m., engineers swung the heavy wooden propeller to start the Rolls-Royce Falcon engine. It kicked into life immediately, and soon warmed into its familiar throaty boom. It was throttled to higher and then lower speeds to check that all was running smoothly, then the ground crew stepped away.

At 4.15 p.m. the two Martinsyde airmen finished saying their goodbyes. Morgan, still with his mischievous smile and twinkling eyes, shook

The "Ramor" Ready To Start, The Propeller Invisible At FullSpeed,   St. John's N.F.

The Martinsyde Raymor ready for take-off.

the hands of some young children who had been brought to the front of the pack. Then the pair climbed into the aircraft, with Morgan taking the navigator's position up front and Raynham occupying the pilot's seat in the rear. They were wearing Royal Air Force flying suits; insulated and warm, but without the waterproofing and flotation pockets that were a feature of the Sopwith outfits. The men settled into their seats. The engine roared in front of them. The crowd edged forward for a final glimpse of the machine. Finally, at 4.21 p.m., Raynham gestured for the chocks to be pulled away. The right hand one came away readily, but the left-hand rope snapped, leaving its chock firmly wedged underneath the wheel. A mechanic approached just inches behind the spinning propeller, grabbed the chock, wrestled it from the wheel, then backed quickly away. Raynham gunned the Martinsyde's throttle and the heavy aircraft started to roll forward. It was an hour since Hawker had taken off.

Like the Sopwith, the Martinsyde bumped and hopped across uneven ground, clearing a path through the gathered spectators as they shouted and cheered their encouragement. With Raynham adjusting the controls as the field's camber threatened to tip the craft from its course, they trundled along the field, picking up speed as they reached the halfway point and continued. But the crosswind seemed destined to keep the aircraft on the ground. In better conditions, Raynham could have taken off by now. Instead, the aircraft was labouring across the grass. Raynham found himself bearing down towards the end of the field. Having travelled almost 200 yards from where he started the run, he still didn't have quite enough lift to get airborne.

He was now travelling at 45 miles per hour towards the edge of the aerodrome. This area had not been prepared for the aircraft's wheels to run over it at such speed. It looked as if they would run out of space. Just then, the aircraft struck a ridge across its path. Raynham and Morgan felt a thump under their seats. Suddenly, the laden aircraft started to crumple to the ground. Raynham, acting instinctively, pulled the control stick back, hard, to lift the biplane into the air. For a moment, it rose. Then, tilting sideways in a gust of crosswind, it crashed back down onto the grass. The weight of the engine and fuel tipped the nose forward. The heavy wooden propeller, spinning furiously, bit into the rocky earth with a loud grating noise. Momentum carried the aeroplane forward for ten more feet before finally it dragged to a halt.

There was a moment of awful silence. Then the crowd began to surge forward. Those closest to the Raymor were shocked by what they saw. It had been wrecked. Its propeller was smashed into splinters. The fuselage, struts, lower wings, wheels, and undercarriage had all been badly damaged. The engine and radiator had taken a beating as the aircraft ploughed into the ground. What was worse, the large aluminium fuel tank had split open on impact. Gallons of petrol were washing over the wreckage.

The spectators who reached the scene first could see that Raynham and Morgan were not moving. One onlooker climbed up the side of the fuselage and grabbed Morgan, pulling him from his seat and down to the ground. Then Raynham, appearing to rouse himself from the shock of impact, climbed out of his seat unaided and dropped down as well. Both airmen were covered with blood.

Realizing that petrol was coursing out of the damaged fuel tank and filling the air with fumes, officials close to the wreckage shouted at bystanders to put out their pipes and cigarettes. It would take only a spark to ignite the whole scene. In the instant after the aircraft crashed, Raynham automatically reached out to switch off the engine magneto, preventing an almost certain conflagration. But some of the smokers in the crowd, never having witnessed an air crash, didn't realize the danger they were in. Some refused to stop smoking. Their cigarettes were grabbed from their mouths and hands by officials, leading to brief fistfights.

Morgan staggered beside the aircraft, bleeding and concussed. Then Raynham was seen suddenly to collapse to the ground. Shouts went up for a doctor. Two were close by in the crowd, as was a nurse. They ran forward to help. By now, Raynham had managed to get back to his feet. He and Morgan were helped away from the wreck towards the road that adjoined the aerodrome. Raynham was able to walk, but he was still in a daze. A few yards on, he shook off his helper's grip and stumbled back towards the aeroplane. Blood streamed from his nose and mouth and an expression of angry frustration contorted his face. Morgan couldn't walk and had to be carried to the road. The first thing he said was "Is Raynham alright?" By then, Raynham had turned once more from the wreckage and was heading towards the road as well, telling nearby friends, "I was just beginning to get a lift. . . . Another twenty-five yards and we would have been away."

At the roadway, the two men were given first aid by the nurse and doctors, who stanched the blood from their lacerations. Morgan was heard to say, "Poor old machine," before apologizing for what had happened. Then they were driven away in two separate motor cars, each with one of the doctors on board. Morgan was taken to the house of a friend with whom he had recently been staying. Raynham was returned to the Cochrane Hotel.

Most of the people in the ten-thousand-strong crowd at Pleasantville were spared a close look at the scene as it unfolded. For those nearby, it was a disturbing experience. The aircraft, so powerful and perfect just moments earlier, was now lying wrecked in a gouged heap of petrol-soaked dirt. Its occupants, two men that St. John's citizens had taken to their hearts, had been carried away injured and bloodied, perhaps worse. Many of the closest onlookers were in tears.

At the Cochrane Hotel, the doctor found to his relief that Raynham's injuries were not as severe as they first appeared. In the collision, his face hit the back of Morgan's seat, causing heavy bleeding from his nose and lip, which soon stopped. Bruising began to appear across his face. Worse was a heavy blow to his abdomen from the aircraft controls as he pitched forward, though the discomfort from this, too, began to subside. After cleaning himself up and drinking some tea, Raynham left the hotel and drove back to the aerodrome to join the rest of the Martinsyde team as they got the aeroplane under cover and figured out what to do next.

At Morgan's temporary residence, the news was not so good. Morgan had passed out for a time during the car journey. Having come to, he was in severe pain. In the crash, his face whipped into the side of the aircraft. His left cheek had been ripped open, needing stitches to close it. His left shoulder was partially dislocated, and the stump of his right leg was badly bruised from where his prosthesis had jammed in the impact. His abdomen had been struck as well. Worst of all was a laceration over his left eye which, for the moment, blinded him. During the afternoon, his pain got steadily worse. In the early evening it was so bad that the doctor injected morphine, but for several hours it had no effect. Eventually, Morgan fell into a broken sleep.

Meanwhile, the Martinsyde team out at Pleasantville surveyed what was left of the Raymor. They concluded that when it hit the ridge at the end of the field, the weight of the machine and the shock from the impact caused the axle to buckle and the undercarriage to collapse. Feeling the collapse and attempting to take off, Raynham only increased the severity of the impact when the machine fell back down, not having enough lift to get away. The sideways gust that tilted the aircraft at that moment compounded the disaster. Later, Ted Brown said the Martinsyde aerodrome was "altogether unsuitable for a 'take-off' into the then wind." Asked why the ridge had not been flattened, one of the Martinsyde men replied that nobody ever thought Raynham would still be running over the ground that far from the start. With a headwind, the aircraft would have risen much earlier, even with its heavy fuel load. But the crosswind, which had made Raynham nervous all day, had done for him just as he feared.

With the gloomy post-mortem set aside, the Martinsyde team's attention turned to the future. There were spares for most of the damaged parts, and enough experts on hand to carry out repairs quickly. But they had only brought one engine, and its repair would likely be beyond the limited workshop capabilities at Pleasantville. Even if they managed to get it working, nobody would trust it to work reliably over the Atlantic. They immediately wired England for a replacement, but it would take two or three weeks to arrive. Worse was the loss of the aircraft's navigator and co-pilot, Charles Morgan. Perhaps he would recover quickly, but for now, he was out of the game. That meant Martinsyde seemed to be as well.

Morgan had a bad night, and he wasn't much improved by the next morning. In the afternoon, he told the doctor and his hosts that he was feeling a little better, though still in pain and blinded in the lacerated eye. Raynham spent the day in his bed at the Cochrane, uncomfortable but healing. He told reporters that he hoped the Martinsyde aircraft could be repaired ready for a second attempt on the contest, but they were sceptical. One of them wrote, "This hope will die when he is again able to leave his room and see what is left of the machine in which he had placed so much faith."

As they recovered from the ordeal of the previous afternoon, the attention of the Martinsyde team, as for all the aviators on Newfoundland, was absorbed by the wait for news of Harry Hawker and Mac Grieve. As it transpired, there hadn't been any need to hurry away from St. John's to beat the Americans. Further reports from the Azores revealed that only one of the Nancies had flown to the islands safely. Navigational problems caused by poor weather forced the crews of NC1 and NC3 to land their machines on the open ocean, miles before reaching the Azores. Neither of them could take off again. NC1 taxied around for five hours before it was spotted by a ship and its crew was rescued. An attempt was made to tow the aircraft to safety, but the line broke and the machine later sank. NC3, after its forced landing, headed for the Azores on the surface, hoping its powerful engines would carry it on the 200-mile journey as a boat. NC4, commanded by Albert Read, was the only craft to have arrived at the islands safely by air that Saturday, and Read was showing no signs of leaving for Lisbon any time soon.

On the Sunday afternoon, when the Sopwith and Martinsyde teams taxied along their respective fields, they were heading into a race among themselves and the American seaplanes. With Martinsyde's disastrous crash and the grounding of the American crews at the Azores, the only aviators left in the race by Monday the 19th were Harry Hawker and Mac Grieve.

CHAPTER 12

NEVER GIVE UP HOPE

IT WAS AT TEN O'CLOCK ON THE EVENING OF MONDAY, MAY 19, twenty-seven hours after her husband set off from Newfoundland, that Muriel Hawker finally received word from the Admiralty. It was the news she had been aching to hear. Harry Hawker and Mac Grieve had landed safely in the sea a few miles off Ireland's west coast, and a small flotilla of destroyers, minesweepers, motor launches, and other vessels was on the way to pick them up. They had not quite made it to the Irish mainland, but they had completed what many the previous day concluded was an impossible journey. The Atlantic, Muriel was told, had been crossed.

It was a joyous moment for the whole watching world. "If, as now seems apparent," wrote the *New York Times*, "they fell short by only a few miles, their feat is hardly less remarkable than if they had actually reached Ireland."

The day had been unsettling. That afternoon, Muriel heard a rumour that the Sopwith aeroplane had been sighted. She drove quickly to Brooklands, where it was thought Hawker and Grieve might go after reaching Ireland. There had been no further news, but it had been good to leave the house. Speaking to a reporter at the aerodrome, she admitted, "I have come here to get away from the telephone. It has been going all day, with all kinds of rumours—that my husband is half way across, or 400 miles off Ireland, and so on. So I just came here to have something to do." When it was apparent that nobody was flying into Brooklands that evening, Muriel returned home. She told a journalist waiting outside the house, "Perhaps I shall not hear of him to-night. He may land anywhere and it is possible that he may not be able to communicate. But I feel confident, even though the news is late in coming, that everything will turn out all right."

About an hour later, the Admiralty message came through, confirming the rumours, leaving Muriel dazed with relief. Before long, as the news began to spread, the telephone started to ring with calls from delighted well-wishers. She shared their joy. But after a while, she could no longer bring herself to take the calls directly. Instead, she left her brother, Leonard, who had arranged to stay with her, to field each message. She turned in for the night, happy and relaxed.

The next morning, Tuesday the 20th, she rose early, went downstairs, and collected up the newspapers from the doormat. She turned first to the *Daily Mail*, where she expected to find enthusiastic reports of Hawker and Grieve's successful crossing of the Atlantic. The paper would be announcing the winners of a prize it had first mounted six years earlier, when the world had been so different. It was a feat of skill and endurance that could now be written into the history books, at a time in history when good news was in short supply.

Instead, as Muriel Hawker opened the newspaper and looked at its headlines, she felt sickened with fright.

Later, she recalled, "I do not think I have ever felt so frantic and yet so completely hopeless as when I saw the fatal words, 'Hawker Missing—False Report of Fall in the Sea.'"

There had been no sighting of the aeroplane off the Irish coast the previous day. Instead, there had been a series of misinterpretations of a wireless message that simply asked ships to look out for the aircraft. It had all been a misunderstanding.

There could not have been a crueller end to Muriel's happiness. She later wrote, "I believe at that moment I gave up all hope."

At ten o'clock that morning, Tom Sopwith called to see her. There was little he could say except to urge her to have faith in her husband and in the aeroplane they had designed. By then, Muriel had taken time to reflect. She recalled the last words that Harry had said before they parted in London, two months earlier. "If things don't go quite right," he told her, "never give up hope." With such devastating reports in the newspapers in front of her, it was a challenge that would be hard to meet. But it was all she could do.

Back in Newfoundland, the mood among the remaining teams had darkened over the course of Monday the 19th, as news of a successful landing failed to materialize. They all knew that Hawker and Grieve had fuel enough for twenty-four hours of flying and no more. After that period had elapsed, the thought on everyone's mind was described by the *New York Times* reporter waiting with the aviators: "It is over. Either they have landed or—"

Fred Raynham tried to remain optimistic that the Sopwith team would "make good," as he told reporters. Others speculated that the pair might have had to ditch in the ocean and deploy the special lifeboat that formed part of the fuselage fairing. The boat was kitted out with two days' worth of food and hot drinks, and the airmen were both wearing US Navy insulated suits. There was still hope, at least for the optimists.

But for the realists, the idea that the pair could survive a ditching in the Atlantic Ocean was laughable. A message on Tuesday the 20th from a correspondent on Ireland's west coast read, "If Hawker and Grieve were in the sea last night with only their life-saving suits on it is almost certain that they were lost. Sheets of rain fell and a gale was blowing. Naval officers say if the airmen could keep afloat in such conditions it would be nothing short of providential."

By Wednesday the 21st, it was clear to the world, as well as to the St. John's aviators, that Hawker and Grieve must have come down somewhere in the ocean. Attention had now shifted from Ireland's west coast, where searches had not revealed any trace of the aircraft, to the waters off Newfoundland. A report from St. John's revealed that several wireless messages had been sent to Hawker and Grieve after they had taken off and were still within range of the transmitting stations. But they had not been answered. In fact, the pair's wireless transmitter, which was new and had a range of 300 miles, had remained entirely silent since the men departed.

A theory was starting to emerge that the Sopwith had got into trouble early in its flight, soon after taking off from the Newfoundland mainland and before its wireless equipment had been deployed. "How else is it," asked reporters stationed at St. John's, "that the voyagers in the Sopwith biplane never sent back a single spark of farewell to the city in which they had been so long, to the technical experts who had helped them so faithfully, or to the rivals with whom they had dwelt in such friendly goodfellowship?" The *Daily Mail* contacted the Admiralty, asking if a British warship was available to search the waters around Newfoundland. The reply came back quickly. It was no.

But in the absence of naval assistance, the Admiralty was nonetheless able to offer some measure of optimism. The previous afternoon, Muriel Hawker had driven up to London with her brother in search of news from Admiralty officials. After the meeting, she told a journalist, "I have not by any means given up hope yet. I thought all Atlantic vessels had to carry wireless, but I am relieved to find that there are so many without." A fleet of cod-fishing boats operated in the fertile waters around the island. The hypothesis was that Hawker and Grieve had ditched early in their journey and been picked up by one of these small Newfoundland craft. If that were the case, news might take several days to reach land. "Hope deferred—but still hope," a *Daily Mail* headline offered.

But most of the aviators remaining at St. John's, who knew more than any the risks that Hawker and Grieve faced, had now lost any optimism they may once have held. Their minds took them back to Gustav Hamel's ill-fated attempt to cross the Channel in 1913. "To this day," reported the *New York Times*, "no one knows what happened to Hamel, and the unspoken question which one may read on every face of the airmen's group here is, 'Will we ever have the faintest inkling as to just what fate befell Hawker and Grieve once they disappeared into the sky?'"

This idea had come to the men in the early hours of the morning as they sat, hollow-eyed, in the lounge of the Cochrane Hotel. Yes, perhaps the flight had ended soon after it had begun, and fishermen had picked the pair up. Or, yes, maybe they had got most of the way to Ireland but had ditched in the sea, safely protected by their life jackets and lifeboat. But the third scenario now being discussed was a sudden mechanical failure, leaving no time to glide down to the water and safely ditch on its surface. Instead, there would have been a plunge straight to the floor of the ocean, where the aircraft would have imprisoned the two men and become their grave.

Later that day, the Sopwith mechanics at Mount Pearl began to pack their stuff into crates, ready to return to Liverpool that Friday. Everyone at St. John's realized what this meant. A report cabled to New York that evening began with this line: "The last lingering hope has died hard here."

By Thursday the 22nd, most commentators had accepted the reality of the situation. An editorial in *Flight*, the weekly journal for aviators, said that Hawker and Grieve's decision to take off, despite reports of high winds along their route, "was a gamble with their luck—and luck failed them." But there could also be discerned a renewed sense of positivity among some. The previous day, the Norwegian navigator for the Handley Page team, Tryggve Gran, raised the possibility that Hawker and Grieve, in trying to avoid storms along their planned route, might have altered their course significantly. Their revised path, he suggested, might follow a course often plied by Scandinavian vessels shipping food from North America to Europe. He told the *Daily Mail*'s St. John's correspondent, "There are at least 50 Scandinavian vessels on this route . . . A large majority of them do not carry wireless, and it is quite possible that Hawker and Grieve may have been picked up by one of them and we shall not hear for two or three days." A columnist writing in the same issue of the newspaper agreed, expressing the hope "that Hawker has glided down near some old tramp steamer or other, been picked up, and will turn up again cheery and bright for another attempt."

Informed though such reasoning was, it was nothing more than speculation. The reality—that the chances of Hawker and Grieve suddenly reappearing had been reduced virtually to zero—was apparent to all but the most doggedly determined optimist.

The following day, Friday the 23rd, the *Daily Mail* announced that it would pay a total of £10,000—equal in value to the transatlantic flight prize, which remained open—to Muriel Hawker and to Mac Grieve's family, once it had been confirmed that the two airmen had been lost. Lord Northcliffe had met Muriel that Wednesday to offer help, and had been impressed by the fortitude she showed during their conversation.

Later on Friday, rumours that Hawker and Grieve had been found, which had spread like wildfire across London the previous evening, were reported in the press. Though the rumours were dismissed, they nonetheless added to the disquiet felt by the airmen's anxious families and friends. On Saturday the 24th, a handwritten note in a bottle, purporting to be a distress message from Hawker, washed up off the coast near Washington, DC. This, too, was quickly deemed a hoax.

But Friday also saw the emergence of more credible sightings. Reports started to surface that day of observations, made in the early hours of Monday morning by ships in the mid-Atlantic, of what had looked like the red light of an aeroplane passing overhead. The times and locations matched Hawker's expected route perfectly.

In a week when rumours, exaggerations, hoaxes, and misunderstandings were flowing around the globe, it was difficult for the Sopwith team to evaluate this latest set of reports. One aviator later wrote, "It seemed that conceivably it might have been a meteor, or even pure fancy." But things seemed to be adding up. "The place where he is supposed to have been seen," a Sopwith representative told reporters, "is the exact spot where he ought to have been." The Sopwith firm concluded that these reports were likely to be reliable and that the red light had almost certainly been the exhaust flames of Hawker's engine.

The sightings could have offered a glimmer of hope to those who believed Hawker and Grieve had been able to glide down to a sea landing before being picked up by a passing vessel. It had taken four days for the ships' reports, made on Monday morning, to reach land. Some ships that had been plying the Atlantic earlier that week had yet to arrive at their destinations. There was still a chance that a steamer lacking a wireless set was at that moment making its final approach to a European port with news of Hawker and Grieve's rescue.

But optimistic assessments like these were tempered by more sobering reports off the arriving steamships. They told of exceptionally severe storms in the Atlantic the previous Sunday night and Monday. A crew member of one of the ships described "wind of great violence and mountainous seas." A Sopwith representative said, "The position of this storm was known to Mr. Hawker before he started, but not its extent or its violence." While still offering the possibility that Hawker and Grieve might show up at a European port in a day or two, the representative admitted it was probable that the storm had claimed them.

If the information emerging from the Atlantic steamers was accurate, then Hawker and Grieve had managed to fly successfully for ten hours and passed the halfway point of their journey, before driving straight into a fearsome and probably inescapable storm.

But the reports might have been mistaken. The only people who really knew what happened in the Sopwith cockpit, after the aircraft passed out of sight of St. John's, were Hawker and Grieve themselves.

CHAPTER 13

NOT YET FAVORABLE—BUT POSSIBLE

HARRY HAWKER AND MAC GRIEVE WERE ONLY TEN MINUTES INTO their journey when they hit a bank of the thick fog that often appeared off Newfoundland's coastline. They thought it a shame to leave behind so quickly the bright, clear conditions they had enjoyed over Mount Pearl that evening. But the arrival of fog didn't worry them unduly. From their observations over the previous two months, they had discovered that these banks rarely extended far, nor were they usually particularly high or dense. More troubling was the discovery that their wireless transmitter didn't work. Because of the way the air was flowing around the aeroplane, the little windmill generator hinged on the side of the fuselage wasn't spinning fast enough to power it.

As they approached the fog bank, Hawker continued to push the Sopwith higher. The Rolls-Royce engine responded enthusiastically to his commands. Before long, still climbing, the aircraft passed through the worst of the cloud. Though they could no longer see the ocean below them, the air above their heads was clear. Ahead, at least for the next few hours, they expected good weather.

Once the aeroplane had reached its cruising altitude of 10,000 feet, Hawker levelled off. The conditions at this height seemed about as perfect as any aviator could hope for. There was no turbulence. The aeroplane, as it pushed along at 105 miles per hour, was so stable it might as well have been flying itself. The layer of cloud beneath them was as flat and uniform as a millpond, the perfect conditions for astronomical observations which could use the flat cloud like a replica of the Earth's horizon itself.

The pair flew on like this for four hours. Warmed by their insulated flying suits and feeling relaxed in the well-equipped cockpit, they managed to keep up a shouted conversation as they took in the sights of their flight. At other times, they cruised along in silence, enjoying the sublime experience of an aerial voyage over this great ocean.

At about 10 p.m., Hawker observed that the cloud layer beneath the aircraft had thickened and changed in form. It no longer appeared to be a milky, uniform ocean of vapour, so flat one could imagine alighting on it. Now it had turned a dull grey colour as the sky above darkened from bright blue to purple. It was patchier as well. Boulders and fissures marked its upper surface as turbulent currents of air contorted the once-flat layers.

Fifteen minutes later, the weather conditions worsened markedly. As they motored along, Hawker and Grieve realized that the fog and cloud around them was thickening fast. They could no longer see anything below them. Then, with their visibility reducing by the second, they were suddenly able to discern the outline of a large storm dead ahead. In the fading light, they couldn't tell how extensive the patch of rough weather would be, nor how violently it might treat the aircraft. With ominous-looking clouds now below and all around them, they had no option but to fly on and take their chances. They both hoped for a merciful treatment at the hands of the storm.

Hawker must have reflected on the observation he had made before the pair set off, when he said that the conditions over the Atlantic made their chances "not yet favorable—but possible."

As the airmen reached the edge of the bank of cloud, the aircraft began to rock around, buffeted by the turbulence. Squalls of heavy rain dashed the pair as they sat in their open cockpit. Hawker fought to keep control of the machine. Soon, the wind inside the storm reached gale force. It howled in the men's ears and forced the aeroplane from its course. Hawker climbed to 12,000 feet, hoping the extra altitude might lift the pair from the worst wrath of the Atlantic weather. It made little difference.

They battled through the storm for a further forty-five minutes. Still the banks of heavy cloud engulfed the aeroplane, leaping up in great mountains with deep valleys between. Below them, the fog was so dense that Grieve had no chance of spying the ocean surface to make speed and direction readings. The layers of cloud above them prevented him from making astronomical sightings, too, though he tried repeatedly.

Just then, Hawker spotted something on one of his cockpit gauges. The temperature of the cooling water that circulated around the engine's cylinders to keep them from overheating had risen, even though the aircraft was travelling level and its engine wasn't being pushed hard. Hawker opened the shutters that covered the radiator, allowing more cold air to reach it. This should have brought the water temperature down immediately. But it made no difference.

The clouds that the airmen had been flying through were getting heavier. Fierce turbulence threw the aircraft around. The air was freezing cold. Hawker, who was wearing thick flying gloves, was unaffected. But Grieve had to take his gloves off to operate the delicate navigational instruments as he tried to work out their position in the worsening storm. He was suffering increasing numbness in his fingers, a first sign of frostbite.

Looking up again, Hawker could see that the cloud formation was too high to climb over without burning a lot of extra fuel. To do so would be a risky option. It might buy them a little calm in this Atlantic storm, but it might also mean they didn't have enough petrol to complete the journey and reach the safety of Ireland. In any case, climbing would make the engine run even hotter. With the fault in the cooling system, this was another reason why Hawker was reluctant to climb higher. He had no option but to keep going at the same altitude and try to dodge the worst of the weather.

Despite his attempts to keep the engine cool, Hawker could see that the water temperature was still rising, and by 11.30 p.m. he realized that something was seriously wrong with the radiator. He theorized that a filter, which kept debris in the radiator from fouling the water pump, had become blocked with fragments of rust and solder. The normal solution when faced with this problem was to undertake a manoeuvre which, in a storm and above an unforgiving ocean, would be risky. The trick was to throttle the engine right down and then dive. This caused the detritus fouling the filter to be pushed to the sides, allowing water to be pumped freely once more around the engine. It wasn't something Hawker wanted to do. But he had no choice. If he did nothing, the water would start to boil. Once it had all boiled away, the engine would overheat and seize.

Throttling the motor to minimum, Hawker pushed the aircraft into a dive, falling 3,000 feet. Then he opened the throttle again. The Rolls-Royce engine thundered back to speed immediately. Hawker pulled out of the dive and climbed back up to 12,000 feet, then looked at the temperature gauge. To his relief, he saw that it had dropped and was holding steady. The manoeuvre, risky though it was, seemed to have worked.

But it was only a brief respite. As they pressed on through the violent wind and turbulent banks of cloud, the temperature once again began to climb. Soon, it reached dangerous levels. Another nosedive would be needed.

This time, the debris-spreading manoeuvre failed. As Hawker pulled the machine out of its dive and climbed once more to cruising altitude, the thermometer showed an alarmingly rapid rise. For a third time, Hawker throttled down and dived, before restarting the engine and climbing back up. Again, the operation failed. Each time he did this, more precious fuel was burned, leaving Hawker anxious that they would consume more in these manoeuvres than they could spare. And now, the water in the radiator had reached the boiling point.

Hawker reviewed the situation. There was no way he could fix the radiator problem. Something had gone seriously wrong, and the usual solution was no longer working. In any case, they didn't have enough fuel for further attempts. The worst of the storm had now passed them, and the weather conditions were starting to improve. Their best option now, Hawker judged, was to stay at 12,000 feet, above the remaining clouds. He would reduce the engine throttle to the bare minimum, hoping this would keep the water temperature below boiling point. It would slow them down, but at this stage, Hawker and Grieve still believed they could make it across. They just had to hope there would be no more storms along the way.

For a further five hours, Hawker nursed the aircraft along. Occasionally, Grieve caught glimpses of the sky and ocean through gaps in the clouds and took navigational readings, correcting their course as the strong winds continued to blow them off their planned route. All the while, the water temperature continued to creep up and the precious coolant was boiling away. But if they could keep going like this, the pair might just have enough water and fuel left to complete the journey.

Then, directly in their path, they saw a bank of dense, black storm clouds so vast that they loomed thousands of feet above the aircraft in a menacing panorama of twisted, contorted forms. It might as well have been a mountain range blocking their course.

There was no way they would be able to climb over this storm. Even without the faulty coolant system, these clouds were just too tall for the Sopwith. At first, Hawker tried to push the aircraft through them, as he had done when they met the earlier storm. But the magnitude of this one eclipsed anything they had experienced. After just a few moments, Hawker realized that he had to get out fast.

His next tactic was to try to dart around the worst of the clouds, twisting and turning the aircraft. But by doing so, he was again putting the engine under strain. The water temperature began to shoot up.

Soon, as Hawker struggled to pilot the craft through the turbulence and violent winds, the water was boiling hard.

There was only one course of action left. With no way to go higher, or to keep going at the same altitude, Hawker would have to bring the aircraft down below the storm. But he had no way of knowing how far down the tempestuous clouds extended. It could be as far as the surface of the ocean itself. If that was the case, the airmen would only see it when it was too late to pull up. It was a risk they would have to take.

Hawker throttled the engine to minimum one more time, pointed the nose down, and began to descend.

After having dropped from 12,000 to 6,000 feet, the pair discovered that the clouds at this level were even thicker and darker than higher up. The aircraft continued to glide downward. The conditions seemed to worsen.

Suddenly, to Hawker and Grieve's relief, they emerged through the floor of the storm. At an altitude of just 1,000 feet, they could finally see the ocean beneath them again. On the horizon, the sun was coming up, and it was a welcome sight. As he prepared to bring the aircraft out of its long descent, Hawker once more opened the throttle to bring the engine back to life.

This time, there was no response.

As they had come down through the freezing cloud with the engine off and dangerously low levels of water, the valves in the carburetors, which provided the cylinders with fuel, had frozen solid. The aircraft was now hurtling towards the ocean with an engine that refused to run. The sea was roiling in turmoil from the storm above. The aircraft had the wind behind it, increasing its speed and pushing it downward ever faster.

Struggling to be heard above the howling wind and the crashing waves, Hawker shouted to Grieve to operate the manual fuel pump at his feet. If Grieve could force fuel into the carburetors with enough strength, he might unseat the valves and flood the chambers with petrol, allowing the engine to restart. Grieve immediately leaned forward and began to operate the pump furiously. If he couldn't get the engine to start, the aircraft would crash into the ocean in a matter of seconds, and with the wind behind it, the impact would be hard: they would almost certainly be killed instantly. Even if they managed somehow to survive the impact, the aircraft, and its lifeboat fuselage, would doubtless be destroyed. The pair would soon drown.

Hawker watched the ocean come up to greet them. He opened the engine throttle again and again in case Grieve's actions had managed to unseat the valves and let fuel into the cylinders. Still the navigator, his head buried in his knees, forced the pump handle back and forth in a determined attempt to clear the blockage. But there was nothing.

When the aircraft was just twenty feet from the waves, Hawker tapped

Grieve on the head to warn him they were about to crash. He shouted at him to sit back in his seat in the hope that his resulting injuries might be lessened. Both men had done all they could to rescue the situation. It was a little after 6 a.m. on Monday, May 19. Harry Hawker and Mac Grieve had made it in an aeroplane halfway across the Atlantic. But the storm had forced them down.

CHAPTER 14

YOUR SUDDEN AND TRAGIC SORROW

AS EACH DAY OF THE DREADFUL WEEK PASSED BY, MURIEL HAWKER refused to give up hope that her husband and his navigator had survived. Every day, after reading the newspapers, she visited the Admiralty in London with Leonard to see if officials could offer any shred of news that might hint at a resolution. Tom Sopwith was in constant touch to offer what comfort he could. And Muriel often thought of Mac Grieve's family. On Friday, she visited them at their home in the village of Droxford, near Southampton. Accompanied by their daughter and grandchildren, Grieve's parents had been forced to live out the days in anguished expectation, just as Muriel had. They received countless telegrams from members of the public expressing sympathy, but it was, as Grieve's father later observed, "a very trying time." Muriel tried to raise their spirits and encourage them to remain optimistic. But even with this network of support and sympathy, and with her fortitude, she found it hard to keep her own spirits buoyed. Each day, hope faded a little more.

For all the excitement of aerial adventure, this was the reality for those left behind: agony and loss.

In one newspaper, Muriel had seen an optimistic report suggesting it might take a week for news to arrive. This notion comforted her. On Saturday, she felt an unshakeable conviction that Harry was about to be found.

She drove that day to visit a group of friends which included Fred Sigrist, Tom Sopwith's engineering manager. One of the men in the party later recalled what she said when she arrived. "You're looking down in the mouth, boys. Cheer up. We shall have some good news to-morrow. Sunday is my lucky day." But her own local newspaper, the *Surrey Advertiser*, carried a report that morning which said, "If they had come down and been picked up by a boat not equipped with wireless, it is certain that the vessel which rescued them would have put itself in the way of other boats possessing the means of sending the news for which it would be known the world was waiting."

After hearing of the *Daily Mail*'s offer of a next-of-kin fund for the Hawker and Grieve families, Muriel Hawker wrote a letter to the newspaper's editor. In it, she said, "Whenever the time comes for my trouble to be relieved, among my happiest duties will be that of teaching our little Pamela that her father did not hesitate to venture all for the honour and glory of the country."

Later that Saturday, Muriel received a telegram from Arthur Bigge, private secretary to King George V. The note read, "The King, fearing the worst must now be realised regarding the fate of your husband, wishes to express his deep sympathy and that of the Queen in your sudden and tragic sorrow. His Majesty feels that the nation lost one of its most able and daring pilots, and that he sacrificed his life for the fame and honour of British flying."

The following morning, Muriel once more gathered up the early editions of the newspapers. For the seventh day in a row, she closely examined each transatlantic flight report. She was still hoping for a glimmer of good news, or at least some suggestion that hope was not finally lost. For six days, there had appeared to be a chance that her husband and Grieve had survived. But after a week with no authentic sighting of the pair or their aircraft after the first night of their journey, it appeared that this chance had now faded to nothing.

The *Sunday Times* carried two reports on Hawker and Grieve's fate. The first was written by Charles Grey, the aeronautical journalist who had grown close to Britain's aviators since the days they had first gathered and flown at Brooklands. "Although there is still just a hope that Mr. Hawker and Commander Grieve may still be alive, having been picked up by a tramp steamer without wireless," Grey wrote, "one fears that we must all make up our minds to the fact that they have been lost at sea." The second article extinguished even that tiny spark of optimism. Its headline simply read, "Hawker search abandoned."

It was over.

FOR THE AVIATORS IN THE VICKERS AND HANDLEY PAGE CAMPS ON Newfoundland, work that week continued under a black cloud. On Sunday the 18th, the day that Hawker and Grieve got away, Jack Alcock and Ted Brown drove down to Ferryland, a town on the east coast of the Avalon Peninsula about forty miles south of St. John's. It was a scouting trip in their attempt to locate an aerodrome site for the Vickers machine. But it proved to be another fruitless excursion. As they drove back to the city that evening, their car was hailed by a passing motorist, who called out, "Hawker left this afternoon."

Brown shouted back, "And Raynham?"

The motorist replied, "Machine smashed before he could get it off the ground."

Once they got back to the hotel, Alcock and Brown discovered what had been going on while they were away searching in Ferryland. Raynham told them what had happened to him and Morgan, and everybody said they were hoping for reports of the Sopwith attempt. Brown could sense what he later described as the "excitement and anxiety" that the world was feeling about Hawker and Grieve's prospects. He said, "Nowhere was it more intense than among us at the Cochrane Hotel, who had shared their hopes and discussed their plans."

But as the week went on without news of the Sopwith's arrival, the residents at the Cochrane became, as Brown observed, "a gloomy crowd indeed." The optimistic spirit had been knocked out of them. They were not set back in their work, nor in their desire to compete for the transatlantic prize. But an atmosphere of happy adventure had turned to solemnity. They had been reminded, in the cruellest way, that the journey facing each of them might lead to the grave.

There had been a positive outcome from the events of Sunday afternoon. After spending Monday recovering in bed, Raynham rose on the Tuesday morning and drove to Pleasantville to inspect his damaged aeroplane again. After conferring with others in the Martinsyde team, he made his way back to the city and stopped at the Blue Puttee dance hall, close to the Cochrane Hotel. There, he hired hall space large enough to act as a temporary repair shop. Then he organized trucks to collect the remains of the aircraft from Pleasantville and bring them to the hall, accompanying the lorries himself on the journey. Once the Raymor was safely at the new workshop, the Martinsyde engineers got straight to work fixing it up. They were not acting alone. Montague Fenn, manager of the Sopwith operation, instructed his mechanics to offer their services to Martinsyde. Jack Alcock offered the Vickers mechanics to Raynham's team as well, at least until the Vimy arrived.

The Pleasantville site was now lying vacant until Martinsyde got their aeroplane rebuilt. Alcock had started negotiating with the Handley Page team to share their Harbour Grace aerodrome, but the discussions were not going well so far. With the Vickers machine due to arrive any day, and without a suitable site having yet been found for it, Raynham told Alcock he could use Pleasantville. Then Fenn, who was starting to pack up the Sopwith operation, told him Mount Pearl would be available if he wanted to take out a lease there as an alternative. Neither location would be big enough for the take-off. The Vimy would need a 500-yard run when full of fuel. But at least the firm's engineers would be able to start assembling the aircraft once it arrived. Alcock would have more time to find a spot suitable for the final departure.

Mount Pearl would have suited Alcock's needs better, but when he made enquiries, he was shocked to learn what the terms would be. The site's owner told him he wanted $2,000 to rent the site until June 15, and $250 each day thereafter. Such rents were unheard of, and Alcock was furious. He needed a site, but he refused to be gouged for a spot of boggy ground covered in stones and ruts. Instead, he gratefully accepted his

Mechanics rebuilding the Martinsyde aeroplane at the Blue Puttee dance hall.

old friend Raynham's offer. The Vimy, when it arrived, would be taken straight to Pleasantville.

That Thursday, as the mechanics and engineers worked on the Martinsyde aircraft at the Blue Puttee, the firm's injured navigator, Charles Morgan, received news from his doctors. He was still suffering the effects of shock, and his facial injuries were severe. He had not recovered the sight in his lacerated eye. It was feared that the blindness in it might be permanent, and it needed to be operated on urgently. His abdomen had also been wounded and it, too, required treatment. His injuries were such that he couldn't risk the stresses of another flight, and he needed to return to England for medical attention. His race was over. At the end of a rueful letter that he addressed to the people of Newfoundland, Morgan wrote, "The doctors have rung the death knell on my ever flying again, but the Raymor will be built anew and guided by Raynham again, and perhaps a better man than myself. You will all be blessed in seeing her soar away above your heads on her great adventure." With that, he bowed out of the competition and began making plans to sail back home to England.

Out at Harbour Grace, work on the Handley Page aircraft had advanced rapidly since the last of the crates arrived there on Friday,

May 16. On the Sunday, while Mark Kerr and Herbert Brackley watched Hawker and Grieve get away at Mount Pearl, their engineers spent the day struggling in rain and mud to build a tall timber gantry over the aircraft's fuselage. The next day, they fitted the nacelles that would house the engines. On Wednesday, one of the aircraft's most critical structural components, a longeron, was refitted after damage caused during the journey from Britain had been repaired. This was a particularly difficult fix, and it was a relief to get it finished. On Saturday the 24th, mechanics hoisted the heavy fuselage up, using the gantry they had built earlier in the week. The undercarriage was wheeled underneath, and the fuselage body lowered onto it and attached. Soon afterward, the four big Rolls-Royce engines were fitted into the nacelles. After a week of hard, exacting toil, the great aeroplane was starting to take shape.

Progress among the teams could not, however, offset the feeling of loss at the Cochrane that week, as hopes for the two Sopwith airmen diminished into grief. Hawker was their friend and they missed him. They missed his practical jokes. They missed the basins of water he propped over hotel doors, and the silly water fights he would initiate in the corridors late in the evenings, when he would run around, laughing, like a big kid. They missed him throwing bread at them in the dining room, and they missed the times he would let the air out of their motor car tyres, leaving them momentarily stranded but in lightened spirits. Most of all, they missed his friendship, his loyalty, and his unswerving devotion to the advancement of aviation. And they all missed Grieve as well. Only Hawker had known the naval commander before the transatlantic contest came along. But the others had taken immediately to the quiet, dignified older man. He might have been more reserved than the Brooklands boys—"just a shy Scot," as his father described him—but he was kind, steadfast, and his loss broke the hearts of those he left behind.

A COUPLE OF THOUSAND MILES ACROSS THE NORTH ATLANTIC from St. John's sits Lewis, the northernmost of Scotland's Outer Hebrides. It is a 368,000-acre island of peat, moss, and freshwater lakes, separated from the western Scottish mainland by a twenty-five-mile-wide strait called the Minch. The island acts as a long, low breakwater. In ancient times it was forested, but it is now mostly treeless, allowing the merciless Atlantic storms to rage across its flat, marshy surface. A precipitous, jagged mass of 150-foot-high cliffs, with treacherous rocks and islets extending from them a quarter of a mile into the sea, rises to form the island's northern coastline, known as the Butt of Lewis. It is reputed to be Britain's windiest location.

Lewis's situation has long placed it on the route of ships voyaging between Scandinavian ports and the Americas. For these vessels, whether sail-powered or steamers, the Butt of Lewis forms a useful waymark for those heading west into the wide expanse of the ocean. For eastbound craft, it is the first landfall after an Atlantic crossing, before passing over the top of Scotland towards Norway or Denmark.

In 1862, in recognition of Lewis's strategic position on intercontinental trading routes, a lighthouse was built on its northern headland. It is a tower of unpainted red brick rising 120 feet above the already vertiginous cliff edge. With a beam reaching out almost twenty miles into the ocean, the Butt of Lewis light has been a welcome sight to mariners peering at it through drenching spray ever since.

Then, in 1890, the maritime information corporation known simply as Lloyd's secured permission to build an additional structure at the Butt of Lewis. On a bluff overlooking the lighthouse from the south, the firm constructed a small, two-storey wooden building not much bigger than a garden shed, with a tall flagstaff planted in the ground alongside. This new signal station was only just large enough to accommodate two signal officers, who also acted as coastguard lookouts. But it connected the island to a rapidly growing web of maritime communication.

The Lloyd's signal station at the Butt of Lewis formed part of an international network of such stations built in the late nineteenth century. It had been an ambitious scheme. By the outbreak of the Great War, there were thirty-five Lloyd's stations circling the United Kingdom coastline. A further 130—some built by Lloyd's, others locally owned but tied contractually into the firm's network—encircled every continent of the globe except Antarctica. The scheme acted as an intelligence routing service. Any vessel that could get within sight of a Lloyd's station could communicate with it using an internationally recognized system of coded coloured flags. Messages sent from ship to shore could then be passed down a backbone of communication channels operated by Lloyd's to the firm's London headquarters. Originally, these were telegraph lines. From 1902 onwards, the cable network had been augmented by wireless telegraphy. From London, messages could be routed onwards to recipients anywhere in the world as necessary. The system worked just as well in reverse.

With signal stations placed strategically around the world's coastlines, vessels could be tracked as they journeyed the oceans. Ship-owners could monitor their fleets. Charterers, consignees, and merchants could send orders wherever in the world they were trading. Weather reports and alerts of ships in distress could be carried through the network. A single message to Lloyd's could be transmitted to any or every signal station in the world, from where it could be signalled to ships as they passed by or put into port. Conversely, a single ship passing even the most remote coastline could get a message anywhere, to almost anyone. Once it had been received at the station, Lloyd's would take care of the rest. All the mariner needed was a set of flags.

The weather over the Butt of Lewis on Sunday, May 25, 1919, was uncharacteristically fine. Winds were light, and the sea around the island's coastline was as calm as a rock pool. Working the morning watch at the Lloyd's station was the chief coastguard officer, William Ingham, and his signal officer, George Harding. The pair expected a quiet shift. But at about 10 a.m., Harding spotted a small tramp steamer making its way around the headland of Lewis from the west.

Tramp ships formed an unsung and often ignored fleet of vessels plying the world's oceans. Without a home port or regular schedule, tramps, unlike the big steam liners of the day, were ad hoc cargo carriers for contract hire. These small but reliable vessels would travel wherever and whenever needed around the world, with voyages often lasting many weeks. Few carried wireless sets. Instead, their owners and charterers knew that these stocky, redoubtable vessels would work the seas slowly but surely once they had picked up their consignment. One industry insider explained, "There is no feverish anxiety as to the hour of their expected arrival. They leave port, pass into the spaces of ocean, and nothing is heard of them till they arrive at their destination, which may be thousands of miles away." Without wireless, communication during an ocean voyage was impossible. But the tramp crews invariably made it through.

As the steamer rounded the Lewis headland and started making its way towards the signal station, neither Ingham nor Harding paid it much notice. Their attention picked up when, at about three miles out from the shore, the vessel began to sound its siren, an unusual act in fine weather. Looking more intently through their telescopes, Ingham and Harding saw that the ship was flying a message with its signal flags. The wind was heading towards land at that moment, pulling the flags end-on to the watching signalmen in their tower and preventing them momentarily from discerning the message. Then, the vessel turned a little and the flags came around. Ingham and Harding saw that the message was the vessel's name: *Mary*.

The pair looked at their list of expected ship movements. *Mary*, a small, steel-hulled steamer owned by the Danish shipping company Christian Andresen, had left New Orleans almost a month earlier. It had taken twenty-seven days to pass 4,500 nautical miles diagonally across the Atlantic. It was now headed for the small Danish seaport of Horsens, on the east of the Jutland peninsula.

As the signalmen watched, *Mary*'s crew pulled down the first set of flags and hoisted a single signal, which the Lloyd's men recognized as "communicate by wire." It was evident that *Mary* had an important message to be transmitted down the telegraph lines to Lloyd's headquarters.

The next flag showed the international code for "saved hands." This was used to alert the recipient that the sender had rescued one or more seamen earlier in their voyage. Then, *Mary*'s signaller hoisted a flag which signified that he was about to spell a message using alphabetical signals. Three flags at a time were then raised. The first set read, "S O P."

These flags were then hauled down and the second set went up: "A E R." These came down and a third set was pulled up, reading "O P L." Finally, the sailor hoisted the letters, "A N E."

Having delivered this cryptic note, *Mary* began to turn away and continue its eastbound journey. Ingham and Harding had quickly worked out what they thought the message meant. But there could be no room for misunderstanding. The pair wanted confirmation before the Danish vessel broke off contact.

"You can judge of our excitement when we realised that the last four hoists read 'Sop. Aeroplane,'" William Ingham told a reporter the following day. "As fast as fingers could bend on the flags we ran up the signal—'Is it,' and the captain put about and came round into signalling range again, while we spelled out in two hoists the word 'Hawker.'"

Ingham said, "Our delight knew no bounds when immediately the reply was run up 'Yes.'"

CHAPTER 15

DON'T CRY

HARRY HAWKER HAD GIVEN UP HOPE THAT THE SOPWITH'S stricken engine would start again. Mac Grieve, as well, had sat up and braced himself for the crash. Both were resigned to a landing that would likely kill them. Then the carburetor valves, which had stuck fast in the cold of their rapid descent, finally and suddenly yielded to Grieve's frantic operation of the fuel pump. As the petrol flooded in and reached the cylinders, the engine fired back to life. "We were within about twenty feet of the ocean when the first gloriously welcome splutter came down the exhaust pipe, and not much above ten feet altitude by the time the twelve cylinders had taken up their duties," Grieve later wrote. "I gave her a good mouthful of throttle," Hawker recalled, "she roared away with the best will in the world, the dive flattened and tilted into a climb and we were soon back again at a four-figure altitude and very glad to be there." He added, "Had we hit the water we should have had not the slightest hope."

But the engine only had enough cooling water left for an hour or two of flight. It was nowhere near enough for them to get to Ireland. The storm continued to rage. Their only hope was to head for the Atlantic shipping lanes and search for a vessel. Then they would have to attract its attention before coming down to land on the furious sea to be rescued.

Hawker flew a zigzag course low over the water, searching intently for any sign of a ship passing beneath. Grieve attempted to take navigational sightings. As the minutes elapsed, the precious water in the engine continued to boil away, escaping as steam from the radiator's relief pipe as an ever-present reminder to the two men of their predicament. Hawker reviewed the level of supplies the pair would have to rely on when he finally had to ditch the aircraft. Neither he nor Grieve had eaten much of their stock of food during the flight. Besides their remaining cockpit rations, they had two days' worth of food and water in the fuselage lifeboat. Hot drinks stored there would keep their body temperatures up. Their life jackets would—they hoped—keep them dry and buoyant. Hawker had packed a device known as a Holmes flare, a large distress signal that would burn brightly for over an hour. He also had several smaller pistol flares that launched a red light high into the sky, though they burned out quickly. Their transatlantic attempt might have failed, but they had a chance, at least, of survival.

"We flew around for some two hours or so in weather that was getting a good deal worse rather than better," Hawker recalled. "There was no lack of rain squalls, the wind was getting stronger and gustier and bumpier every minute, and the sea rougher. . . . As may be imagined, I was not altogether without anxiety although I knew we were right on the steamer route, to which we had taken care to shape our course, because there was plenty of fog, and we might have passed quite close to a ship without seeing her."

Hawker reflected on his decision to fit the aircraft with a detachable undercarriage, which he had jettisoned over the waters off the Newfoundland coastline. It was a tactic that had attracted criticism. Now, faced with the prospect of landing on water, Hawker felt vindicated. With heavy wheels protruding from the fuselage, the aircraft would have dug into the water, catapulted over, and almost certainly killed its occupants. Without the wheels, Hawker might be able to land safely.

As the sun continued to rise on that Monday dawn, the pair flew backward and forward, thrown about by the tempestuous wind and with the ocean crashing beneath them. They were getting ever closer to the moment when the last drops of coolant would boil away and the engine, which had answered their every demand so far, would finally seize. The conditions were among the worst they had experienced during the flight. "It was like being in a small motor-boat in a heavy sea," recalled Grieve. Hawker, who had always suffered from debilitating sickness during turbulent flights, began to throw up repeatedly as he struggled to keep the aircraft on a steady course. Grieve found himself grinning at his pilot's obvious distress.

The weather continued to close in on the aeroplane, cutting off visibility until the pair were almost flying blind. At 8 a.m., two hours after the dive through the storm that nearly killed them, the final reserves of cooling water boiled away. The engine would seize and fail at any moment.

Then, another miracle occurred.

"Suddenly," Hawker remembered, "a hull loomed out of the fog and we knew that our luck, if it had been patchy, was at least good enough to stand up when the big strain came on it." Flying at just 400 feet above the waves, the Sopwith was almost overhead as the ship came into view. Hawker let out a shout of joy and relief as he spotted the vessel. It was the Danish tramp steamer *Mary*.

Despite his continued sickness, Hawker steered the aeroplane in a circle around the vessel, firing off three distress flares. The aircraft was immediately seen by the steamer's second mate, Christian Hoey, and the helmsman, Ova Schwartz, who were on watch at the time. As the airmen circled, they saw other members of *Mary*'s twelve-strong crew arrive on the ship's deck. Once Hawker was sure he had been spotted, he pushed the aircraft forward by two miles along the ship's course, before turning round. Then, flying into the wind, he brought the Sopwith down onto the water.

It was as good a landing as the men could hope for. The smooth under-surface of the machine's fuselage rode easily over the water. The partly emptied fuel tanks offered buoyancy. The loss of cooling water in the engine's cylinder jackets provided further air pockets that would help keep the Sopwith afloat. But the lower wings of the biplane quickly became submerged. As the heavy twelve-foot waves dashed repeatedly against the craft, it began to break apart.

Hawker and Grieve were reassured to see that the steamer was still heading in their direction. But with seawater rapidly filling their open cockpit, the pair had to climb up onto the seats, bracing themselves against the cockpit walls and holding tightly to the struts supporting the upper wings. They reached back over the fuselage, unclipped the lifeboat, flipped it over onto its keel, and launched it into the water, where it was tethered by a line to the wreckage of the aeroplane. For now, it was safer for the two men to remain with the larger structure. The little boat and its provisions stood ready in case the aircraft began to sink.

With *Mary*'s thirty-four-year-old captain, Adolf Duhn, now on the bridge, the steamer closed in on the two airmen. The sea was too high for a quick rescue. When *Mary* had come within 200 yards of the Sopwith wreckage, the Danish crew began to launch the vessel's lifeboat. The rough sea and raging north-easterly gale meant this hazardous operation alone took thirty minutes.

Once the boat had been lowered over the side of the deck into the water, five of *Mary*'s crew descended the rope ladder hanging over the ship's steel hull and climbed aboard. Frederik Schubert, *Mary*'s chief mate, commanded the boat and led the rescue. Second mate Christian Hoey and helmsman Ova Schwartz, who had first spotted the circling aircraft, joined Schubert in the craft. Kristian Larsen, the ship's carpenter, and seaman Fred Jensen completed the crew. With the boat heaving in the water, the five men began to labour with their oars through the swelling waves, paying out a line from *Mary* as they did so.

It took an hour of hard rowing by the Danish sailors to cover the short distance between *Mary* and the Sopwith. Without the life-preserving jackets keeping Hawker and Grieve warm and dry as they waited in the cold water, they would surely have perished. Moreover, had they come down an hour later than they did, the sea conditions would have been too rough for *Mary* even to launch its lifeboat. In these circumstances, too, their lives would have been lost.

Finally, the lifeboat crew reached the wreckage of the aircraft. As the boat was tossed about by the sea, it repeatedly struck the side of the partly submerged aeroplane. Then the Danes hauled the two airmen aboard.

Schubert waved at *Mary*'s deckhands, who began to pull the boat back to the steamer using the line it had run out during the transfer. Once it had been pulled to *Mary*'s side, its occupants were faced with an intimidating climb up the rope ladder to the level of the steamer's deck, which pitched and rolled in the swell.

Hawker and Grieve arrived on board *Mary* exhausted from their ordeal. Hawker's nausea, compounded by the boat transfer, had left him weak. Both men declined the food they were offered by the Danish sailors. Instead, they asked Duhn the likelihood of encountering wireless-equipped ships in this part of the ocean. Duhn told them he would expect to see one at any moment. Relieved, Hawker and Grieve went below to sleep. But the storm that caused the airmen such tribulation continued to intensify, and it didn't treat ships any easier than aeroplanes. For a while, *Mary*'s crew had to slow the vessel almost to a halt as the wind continued to howl around it. The little steamer eventually pushed on. But it had now been separated from the busiest shipping lane. The chances of meeting a vessel fitted with wireless faded.

It took seven days after the rescue for *Mary* to complete its crossing of the Atlantic. During that time, Hawker and Grieve were made comfortable by the steamer's captain and crew. After two days, Hawker's travel sickness abated. The pair ate well and passed the time reading books in English that Duhn, a fluent English speaker, carried on board. The airmen knew that their families would be spending the week in dreadful anxiety. But there was nothing that could be done. The ship did not encounter a single other vessel for the rest of its transatlantic voyage.

ONCE HE DEPARTED FROM THE BUTT OF LEWIS ON THE MORNING of Sunday, May 25, having signalled news of the rescue, Adolf Duhn resumed his journey towards Denmark. Meanwhile, William Ingham and George Harding, the coastguard and signal officers, sent an urgent telegram to the London headquarters of Lloyd's, reporting the messages that had passed between *Mary* and the Lewis headland.

Within an hour, the Admiralty had been informed by Lloyd's of the message. Still smarting with embarrassment from the incorrect report they had issued a week earlier, Admiralty officials were cautious about accepting the news without corroboration. They contacted the naval battle fleet stationed in Orkney's Scapa Flow and instructed officers to despatch fast destroyers to intercept *Mary* as it trudged eastward from Lewis. Only when the report had been verified would official confirmation of the airmen's salvation be issued.

The Scapa Flow commanders were unsure of the route that *Mary* would take as it rounded Scotland's coast and headed for the Jutland peninsula. Most likely was Pentland Firth, the strait that separates Orkney, off the north-eastern tip of Scotland, from the mainland. An alternative was the more northerly Fair Isle Channel, between Orkney and Shetland. Officers sent ships in both directions. After several hours, one of the vessels, *Woolston*, was approaching the inlet of Loch Eriboll, sixty miles west of Scapa, when its lookout spotted *Mary* making its way towards them. The two ships made contact. *Woolston* signalled to *Mary*'s crew that it wanted to take Hawker and Grieve on board so they could be landed on British shores as soon as possible. The steamer could then complete its journey to Denmark without delay. Adolf Duhn agreed.

Harry Hawker and Mac Grieve bade an emotional farewell to Duhn and his crew before transferring to the British destroyer. Duhn later gave an account of the week they had shared with the airmen. "They were pleasant fellows," he said, "with whom we were the best of friends. When they had had their sleep out they got a good meal with a glass of Schnapps and a whiskey, and they were then quite all right again." Hawker later wrote, "Neither Grieve nor myself can possibly find words to express our deep gratitude to Captain Duhn of the good ship *Mary*. His men had extreme difficulty in taking us off, and we owe our lives to their gallantry, for there is no doubt that, as Captain Duhn said, in another hour we should have gone down for keeps."

Tramp steamers carried sailors hardened by years of maritime experience. Tramp crews spent long periods away from land and family and were admired for their resourcefulness and dependability. One shipping journalist wrote, "They are a class apart, these small tramp men, whether British, Norse, or Danish—excellent sailors with the salt in their bones, who go to sea because the sea calls them as it did their fathers before them."

Adolf Duhn, a quietly spoken and thoughtful young man who had been captaining transatlantic steamers since he was twenty-six years old, would later play down the dangerous nature of the rescue his crew carried out. But five of the men under his command risked their lives that morning. All five, in addition to Duhn himself, were later awarded gallantry medals by the British government. The king personally intervened to get the awards rushed through. On hearing of the medals, Duhn said, "I am very pleased that the King of England should pay such a tribute to us for what we did, and for what we merely regard as our simple duty as sailors, about which there was no need to make any fuss at all."

Once Hawker and Grieve were safely boarded onto *Woolston*, *Mary* and its life-saving crew proceeded towards their destination at Horsens, while the destroyer returned to its base in Scapa Flow. There, the airmen were transferred to the flagship *Revenge* to spend the night. They were given what Hawker described as a "splendid welcome home" by the fleet's admiral, Sydney Fremantle, and his crew. Life in a peacetime battle fleet at such a remote location could be a hollow, monotonous existence. Fremantle later observed, "If Scapa Flow presented few attractions in wartime, it had fewer still after the 'Cease Firing' had sounded." The presence of the rescued aviators, even for only a night, gave the ship's 1,200-strong crew a much-needed lift.

The next morning, Monday the 26th, Hawker and Grieve disembarked *Revenge* and were taken by another destroyer to the Scottish mainland port of Thurso to catch a train to Inverness, and, from there, another to London. It was a bright and clear day. As the pair climbed onto the landing stage at the town's pier, a wave of cheers erupted from crowds that had gathered there. The Thurso town provost, Donald Mackay, addressed the airmen and the crowd. He said, "From the moment of your departure from St. John's the world has been on tension for news of you. Expectation gave way to anxiety, and then anxiety to gloom, but happily all fears and forebodings are to-day dispelled."

Hawker and Grieve were treated to a lunch at Mackay's residence. Then they were taken through streets lined with well-wishers to Thurso railway station, where they boarded the 2 p.m. train to Inverness. They expected to arrive in London at about seven o'clock the following evening. At every station along the line, large and small, the airmen were given the sort of reception normally reserved for returning war heroes. Each time the trains came to a halt, Hawker and Grieve were besieged by crowds of residents and local officials desperate for a glimpse of the saved airmen. "Smile, Grieve, smile—as you did when I got sick," Hawker joked as photographers crowded one station platform. Everywhere they turned they were met with cheers of exultation. The final stop before London was to be Grantham, ninety-five miles north of the capital, where Harry would be reunited with Muriel. It was, in the words of one reporter who travelled down with the airmen, a "riot of human emotion."

AFTER READING THE HEADLINES IN THE SUNDAY NEWSPAPERS reporting that the search had been called off, Muriel Hawker walked over to the parish church opposite her house. Inside, the vicar offered prayers for the safe return of the missing airmen. Soon after returning home, Muriel took a telephone call. It was the *Daily Mirror*, bringing her news of the pair's rescue.

The rest of the day passed in a blur of emotion as Muriel and her brother, Leonard, toured the area to share the information. Reports spread quickly. "The telephone was at work all day," said the *Daily Mail* the following morning. "People in motor-cars shouted, 'Is it true?' and waved their hats in recognition of the answering 'Yes.'" Later, each of the two airmen sent a short telegram to their closest family member. To Muriel, Harry wrote, "All well, making for home." Mac Grieve's parents received a note from their son that read simply, "Safe. Will wire later."

Back at the Hawkers' residence in Hook, a crowd of journalists and supporters grew steadily, numbering more than two hundred by the end of the afternoon. One of Muriel's neighbours told reporters, "She is a most wonderful woman. How she has stood the awful suspense I

The Hawkers' villa at Hook after Harry's rescue had been announced.

don't know, but she has been totally indifferent to bad news and good, and to look at her now you would not think that only a few minutes ago she had been told that the husband whom everybody else thought dead was alive and well."

One reporter asked Muriel why she had been so sure her husband would turn up on the Sunday. She replied, "I can't remember exactly what made me regard it as my lucky day, but I have done so for a long time. It was on a Sunday that my husband made his highest flight. It was on a Sunday that he started from Newfoundland, and lots of lucky little things, too domestic to be worth telling, have happened to me on a Sunday."

That evening, Muriel and her sister returned to Hook church for a special service of thanksgiving. After it was over, the pair went with a party of thirty to a dinner in central London. Muriel returned to a telegram from the king's secretary, which read, "The King rejoices with you and the nation on the happy rescue of your gallant husband. He trusts that he may be long spared to you." Lord Northcliffe, who had been so moved by Muriel's fortitude when they met earlier in the week, wrote,

"Please let me express my delight at the accuracy of your prediction of Wednesday last."

The next day, as Harry was travelling south from Thurso, Muriel attended the screening of a film showing footage of her husband's preparations at Newfoundland. Accompanied to the cinema by her mother, brother, sister, and Fred Sigrist from the Sopwith company, Muriel was left reeling by the cheering crowds she described as "simply mobs of people."

On Tuesday, Tom Sopwith accompanied Muriel to Grantham station, where the stationmaster arranged for Harry and her to meet in his little office on the platform.

Speaking in Newcastle during the journey, Harry said, "I have been loyally backed up by my wife, and when a man embarks on an adventure of this kind the spirit in which it is taken by his wife counts for a great deal. She has been splendid through it all, and what credit there is for what has been achieved is hers as much as mine."

Eventually, the train bearing Hawker and Grieve to London pulled into Grantham. Harry stepped out from his compartment and was ushered by railway officials through the crowds to the stationmaster's office, where Muriel was waiting. She remembered, "He just said the sweetest and most wonderful thing I could ever hear, and added, just as the people started to crush in, 'Don't cry.'"

Vast crowds met the aviators off their train once it arrived at King's Cross station in London. "They stepped into a human cauldron, boiling with enthusiasm, seething with demonstrative energy," wrote a *Daily Mail* reporter on the scene.

A reception at the Royal Aero Club was followed by an open-air concert at the Sopwith aircraft factory at Ham. Dinner with Tom Sopwith at a restaurant near his firm's Kingston works came next. The Hawkers finally arrived at Hook just before midnight.

As the pair drove through the village towards their house, they saw a large display of multicoloured lights spelling out a welcome. The couple's daughter had been kept awake by the nursemaid. "Mr. Hawker stepped from the car and took little Pamela from the nurse's arms," reported the *Daily Mail*. "The crowd's cheers were frantic."

Mac Grieve, Tom Sopwith, Muriel Hawker, and Harry Hawker photographed at Grantham station before rejoining the train and heading for London.

The exultation of the crowd at Hook that evening knew no end. At the thanksgiving service in the parish church that afternoon, the vicar's reading had been from the gospel of St. Luke: "For this my son was dead, and is alive again; he was lost, and is found."

Harry Hawker's transatlantic attempt had ended in failure. But he was alive. He was reunited with his beloved wife and child. After all the trials he had endured, he was home.

Meanwhile, Mac Grieve returned to his family home in the picturesque village of Droxford a couple of days later. Accompanied by his father and brother, he stepped off the train to find the station adorned with colourful bunting. His mother and niece were there to greet him, showering him with kisses. St. George flags had been flown at churches along the way. Laburnum and chestnut trees were in full, fragrant bloom, filling the village air with a heady, warm scent. Hedgerows were dappled with white and red blossoms. A large party of schoolchildren lined the road outside the station. As the Grieves walked by, the children sang a song written by a local schoolgirl, before the village rector gave a welcome address. Then the family climbed into a waiting carriage which was pulled through the village by a group of local workers and service personnel to the loud strains of a band playing triumphal marches. Boy Scouts, firefighters, more schoolchildren, and villagers crowded the road towards the village green, where the carriage halted. To lusty cheers, a welcome-home speech was given: "As one family we have come here to-day to welcome one of the family home. We welcome him with pride and gratitude. We are proud that the sailor son of a sailor father has added fresh honour and lustre to the traditions of a glorious service."

THE NEW YORK *GLOBE* PRINTED A STIRRING ASSESSMENT OF THE flight and what it meant for the world. Describing the endeavour as "sublime insanity," it read, "these two young men, Hawker and Grieve, have done better than make a sober addition to the science of flying— they have given us a lesson in the art of living and dying. How pitiful our daily precautions, our comfortable provisions against penury and old age, our damnable prudence, in the light of such a spirit as was in these men. They have not glorified a country; they have enriched mankind."

CHAPTER 16

MADDENING

IN NEWFOUNDLAND, IT WAS THE NEWS THAT EVERYONE HOPED for, but nobody thought would come. On the afternoon of Sunday the 25th, it took less than an hour for reports of Hawker and Grieve's rescue to spread across St. John's. The public reaction was one of exuberant celebration. Flags that had been lowered to half-mast earlier in the week were raised again. Every notice board was crowded by citizens looking for updates. Every church shared the good news at its evening services.

Among the aviators on the island, the response was initially sceptical. The airmen at the Cochrane Hotel recalled only too painfully the situation the previous Monday, when positive reports had soon been found wrong. At first, the news that came through on Sunday was dismissed as well. But as the day progressed, urgent requests for verification brought back positive responses. It gradually became apparent that this time, the reports were true. "It was in men's eyes rather than in their conversation that there was to be gained some idea of what the eleventh hour message meant to them," wrote the *New York Times* correspondent at the Cochrane. The black cloud of sorrow that had settled over the hotel began to disperse.

The two Sopwith men had left an indelible mark on the other aviators at Newfoundland. Mark Kerr later said, "It always did one good to meet Hawker—he was like a cocktail when one felt down on one's luck." Ted Brown observed that, "For Grieve, as an expert on aerial navigation, I have the deepest respect." For the transatlantic contestants who remained on Newfoundland, fire had been kindled in their hearts once more.

Jack Alcock had spent the day at St. John's Harbour. The previous evening, the steamship *Glendevon*—beset by delays caused by storms and ice—had finally docked with the Vimy aeroplane and more Vickers mechanics on board. Under Alcock's supervision, the crates, cases, and drums of fuel were unloaded from the steamer's hold and placed on the dock. Charles Lester and his men were again pressed into service. By the end of the day, most of the crates had been loaded onto Lester's trucks ready to move. The following day, the transfer was completed. The trucks were transported to Pleasantville, where the newly enlarged Vickers team set to work unpacking them.

Alcock set a tight work programme, carefully scheduling tasks so that the limited time would be used effectively. A setback came when they discovered that the fuel Vickers had shipped from England, all twenty-three drums of it, had spoiled. It had taken on a viscous quality that they feared would clog the engines. They immediately cabled the Vickers factory to send a fresh batch, but it would take time to arrive. When he heard about the problem, Fred Raynham immediately offered the men Martinsyde's spare fuel to use while they waited.

Trickier was the fact that the Vimy was too large to fit inside Martinsyde's canvas hangar. Its wingspan, at sixty-eight feet, was only half that of the giant Handley Page bomber, but it was half as large again as the span of the Martinsyde. The Vickers mechanics used the tent mostly as a store and headquarters, although they managed to poke the aeroplane's tail inside. As at Harbour Grace, the main body of the aircraft would be assembled in the open air.

In the days that followed, the citizens of St. John's, having grown familiar with the Pleasantville aerodrome during the weeks when Martinsyde occupied the site, started making their way back there in droves. The Vickers men were working long, trying shifts, labouring for up to fourteen hours each day before heading for their digs to catch some sleep.

With such intense activity requiring concentration for long stretches at a time, they found the attention from the crowds distracting. Screens were put up around the aircraft to give the team privacy. Worse was the behaviour of those visitors who came to inspect the work close-up. They had a tendency to fiddle with the aeroplane. Some would test the tautness of the doped fabric using the tips of their umbrellas, causing damage that needed to be fixed. Others leaned nonchalantly on the delicate edges of the wings—"much as Australian soldiers on leave from France used to lean against the lamp-posts of the Strand," Ted Brown observed. One visitor, possibly inebriated despite the island's prohibition, pressed the tip of his lit cigar against a wing surface to see if it would catch. He seemed a little put out when officials told him to stop.

IN MID-WEEK, NEWS REACHED THE TEAMS THAT THE AMERICAN navy seaplane NC4 had taken off from the Azores early on Tuesday the 27th, bound for Lisbon. It landed safely later that day. The Atlantic had now been crossed by air. It had taken place in hops, with eleven days elapsing between take-off at Newfoundland and landing on the Portuguese coast. But it was a profound moment in history. The new world had been joined to the old by air, albeit slowly. A few days later the final hops—from Lisbon to Figueira, then to Ferrol, then to the southern coast of England—were completed. On May 31, to the cheering of crowds and the shrieking of sirens and whistles, American navy commander Albert Read landed NC4 in the harbour beneath Plymouth Hoe.

IN HARBOUR GRACE, 2,000 MILES FROM PLYMOUTH, WORK AT THE Handley Page aerodrome had been proceeding steadily. With the engines fixed safely in position, the next part of the assembly was to fit the aircraft's wings. The Handley Page biplane had wing surfaces that covered an area of 2,800 square feet, making their manipulation and attachment a difficult task. Moreover, a complex matrix of struts and tensioning wires needed to be installed as part of the job. These held the wings in place and kept the upper and lower planes in the correct configuration. Fitting all these interconnected items together in the calm conditions of a factory was enough of a challenge. Doing so in the open air would test the mechanics to their limits.

On Friday, May 30, just when the Handley Page technicians planned to attach the wings, a vicious storm broke over Newfoundland. Bitterly cold rain, downfalls of snow, and gale-force north-easterly winds began to pound the Avalon Peninsula. Fog rolled in from the coast. The work being carried out at Harbour Grace became perilous. Realizing how risky the wing attachment would be in such conditions, the men set down their tools. Then they hammered a line of telegraph poles into the ground to support canvas strips forming a windbreak. With the modest protection the barrier provided, the wing assembly was restarted. After two days of hard slog, it was eventually completed. As the storm continued to rage, further work on the aircraft was halted. Mechanics turned the machine so that it faced into the wind, lashed it to the ground, and retreated while the storm blew itself out.

At the Vickers site in Pleasantville, the storm came at a similarly crucial time in the assembly process. The Vickers men had only had four days to work on their machine since it arrived on the *Glendevon*. They had made rapid progress. The twin engines had been fitted on Thursday the 29th, and much of the wing attachment took place the following day. But the rising winds and torrential rain became too much. With nothing for it but to lash tarpaulins over the engines and hope for the best, the Vickers mechanics shut down the site and withdrew.

Besides supervising the assembly work at Pleasantville, Alcock had continued the search for a suitable take-off site after suffering disappointments at Mount Pearl, Ferryland, and elsewhere. News had spread that Vickers was still on the hunt for land. An offer of some pasture at Harbour Grace led Alcock to head out and inspect it. Once he got there, he found that the site, for which the owner wanted the exorbitant rent of $25,000, was no bigger than Pleasantville. When he got back to St. John's that evening, his "one-sided grin," as Brown recalled, said it all.

Throughout the search, Alcock continued negotiating with Handley

Page to share the big aerodrome at Harbour Grace. A breakthrough was made on Tuesday the 27th. A Handley Page official finally agreed to the proposal on the condition that Vickers would contribute towards the sum that had been spent preparing the site. This seemed fair. The deal would give the Vickers team an aerodrome big enough for their large aircraft to depart from, and it was good, level ground. It also showed that Handley Page shared the spirit of cooperation that had been evident among the other transatlantic contestants from the start. Alcock got permission from Vickers to pay towards the costs and agreed to the terms. Then he announced to the press that the Handley Page and Vickers aeroplanes would both get away from the same field—and probably at the same time. The *New York Tribune* reported, "Both aircraft, respectively the largest and second largest now making ready for the 'big hop,' will have their first trial flights about June 1."

But the following morning, Alcock received a telegram from Mark Kerr. The admiral was vetoing the agreement. In the message, Kerr informed Alcock that Vickers could use the Harbour Grace aerodrome only after Handley Page had departed on their transatlantic flight. Moreover, Alcock's firm would be expected to pay the entire cost of the site's initial preparation. The deal was off. Worse still, Handley Page was trying to freeze Vickers out of the race.

Alcock had not seen this coming. From the start, there had been healthy competition among the teams. But nobody played rough. Instead, each competitor looked out for the others while pushing their own team forward. They offered help and support when it was needed. When the stress caused by interminable bad weather threatened the safety of the Sopwith and Martinsyde endeavours, Hawker and Raynham agreed to give each other warning of a departure attempt. When Raynham needed help to repair the stricken Martinsyde aircraft, Sopwith and Vickers immediately offered their mechanics to the task. When Alcock needed a temporary assembly field, Raynham gave him Pleasantville and the use of his hangar. When the Vickers fuel was found to be contaminated, Martinsyde offered theirs with no conditions and no hesitation.

The three St. John's teams dined together, they went to the movies together, they played cards together. They mourned together during that terrible week when Hawker and Grieve appeared to be lost.

Hawker, Raynham, and Alcock had been friends since the earliest days of aviation. Grieve, Morgan, and Brown had quickly fallen in with the sporting spirit that existed among their pilots. Brown became renowned for his quiet willingness to help anyone outside the Vickers team who needed it. Yet Vice-Admiral Kerr, who moved in more exalted circles than the young Brooklands men, had shown, with the message that Alcock now held in his hands, that friendship didn't matter to him. Evidently, Kerr wanted Handley Page to win at any cost.

Alcock and Brown were, in the words of one reporter that day, "thoroughly disgusted." They were disgusted by the price-gouging they experienced in Mount Pearl and Harbour Grace, where local landowners saw the opportunity to make a quick buck from poor land. And they were disgusted by Kerr's sudden refusal to share the Handley Page aerodrome.

But Kerr's unexpected message had a galvanizing effect on the Vickers team. "This means a fight," said one of them after Alcock broke the news, "and the glory of getting there first will be all the greater for us now, if we can make it."

Then Alcock's fortunes took an upward turn. Charles Lester, the contractor who carted the Vickers cargo to Pleasantville that Monday, told him about a field he owned a mile or so south-west of downtown St. John's. He asked if the Vickers team might want to use it. Alcock and Brown went out to look at the site. It was a large meadow that Lester used to graze his horses. Some of it was marshy, and a hill occupied half of the site. There was a level patch, though it wouldn't be long enough for the Vimy's take-off. But the site showed promise. The airmen began to explore the land outside the boundary of Lester's plot. They found four adjoining fields which, if they were united into Lester's meadow, would give the men the 500-yard run they needed, albeit with 100 yards on a slope.

Alcock had to bargain hard with the owners of the other fields. On Friday the 30th, after much negotiation, he struck a deal. Lester put thirty men onto the task of taking down walls, removing trees, filling in ditches, carting away boulders, and flattening the ground. By June 8, the new Vickers aerodrome was ready. The location took several names. Some described it as Ropewalk Field, referring to a prominent rope-making factory nearby. Alcock and Brown sometimes called it Monday's Pool, after Mundy's Pond, adjoining the land. Most knew it simply as Lester's Field.

It wasn't until Tuesday, June 3 that the storm over eastern Newfoundland abated and the rain and snow stopped falling. At last, the two aerodrome sites could be reopened. At Harbour Grace that morning, the Handley Page navigator Tryggve Gran told reporters, "Two fine days will see all the work done, and then it will not be long before we are off." The Vickers team returned to Pleasantville, where assembly of the Vimy restarted.

Having arrived only recently, and having lost time to the storm, the Vickers men had a mountain of tasks to catch up on. Under Alcock's affable but determined direction, they worked hard. Each morning, after he had driven the team from their digs out to Pleasantville in the seven-seat Buick, Alcock cooked the men bacon and eggs in a skillet over a twig fire for their breakfast. For lunch, he prepared them sausages and fried tomatoes. Two of the engineers decided to save the time taken up by travelling each day by moving their digs on site, bedding down overnight in the big crates that once held the Vimy's wings. One of them was a man called Bob Dicker. He and Jack Alcock had first met at Brooklands in 1911, when both men had recently turned eighteen. Dicker was a motorcycle fanatic, so when Alcock arrived at Brooklands on two wheels, the pair hit it off right away. By 1919, Dicker was an engineer for Vickers, and was given the job of modifying the Vimy's complicated control systems for the transatlantic hop. He had a great deal of exacting work to complete at Pleasantville. In later life, he recalled the episode fondly, writing, "What a grand lot of lads, never a moan, and the conditions were far from home standards."

The Vickers men raced to complete their work, and on Monday the 9th, the Vimy was ready for a test flight. It was a perfect day. Alcock arrived at Pleasantville at 8 a.m. and began to cook the mechanics their breakfast as they made final preparations and adjustments. As the day wore on, huge crowds of St. John's residents, who had heard rumours of an imminent take-off, started gathering on the airfield to watch. At 4 p.m., the engines were started. For an hour, mechanics gripped the aircraft's wings and tail, holding the machine back as tests were carried out. Alcock was happy. Then he and Brown climbed into the cockpit and took their seats. Alcock took the right-hand position behind the large, round control wheel, while Brown sat on the left, with his sextant and the wireless controls within easy reach. The wind was coming from the west, a good direction for take-off. After settling in, Alcock pronounced that he was ready and shouted the all-clear.

The crowd was treated to a flawless departure. The chocks were pulled away from the wheels and the machine began to roll along the bumpy grass field. With a light load of petrol in the tanks, it quickly picked up speed. The headwind increased lift on the wings. After only 60 yards, the big bomber eased into the air. The sun shone in a bright blue sky. Alcock pushed the aircraft out over the sea. Climbing to 4,000 feet, the airmen looked down to the water, seeing it reflect the deep hue of the sky overhead. There were icebergs along the coast, and the flickering appearance of whitecaps.

For over thirty minutes, Alcock piloted the Vimy effortlessly over the St. John's coastline while Brown made observations with his sextant. Alcock felt at one with the machine, and Brown was happy as well. He later recalled, "Under Alcock's skillful hands the big Vimy became almost as nippy as a single-seater scout." They circled over St. John's for ten minutes or so, bringing more of the city's residents into the streets to watch them. Then the time came for Alcock to bring the Vimy back down to land.

The trip was acting as a positioning flight as well as a test. Before heading for Pleasantville that day, Alcock arranged for a fire of spruce branches to be lit at Lester's Field, and its thick column of smoke was acting as a navigational beacon. The crowds who witnessed the aircraft taking off at Pleasantville had rushed over to Lester's Field. Up to a thousand more people joined them there. Leaving the city's downtown behind,

Alcock flew west, fixed on the smoke signal, and descended. It was a perfect landing. As Alcock taxied the Vimy to a halt, the crowd gave up a great cheer. A reporter observed, "Alcock's always ruddy face took on a deeper red at the unexpected tribute."

After climbing out of the aircraft, Alcock and Brown helped the mechanics push it to a sheltered spot on the field, then fixed it down to the ground and put a rope barrier around it. Addressing journalists, Alcock said, "I am perfectly satisfied with the machine. She behaved splendidly, and I shall be off to England as quickly as I can. I hope to be in London before the week end." At the post office, he cabled Vickers in London to tell them, "Machine absolutely top-hole."

MEANWHILE, FRED RAYNHAM AND HIS TEAM CONTINUED REPAIR-ing the Martinsyde aeroplane at the Blue Puttee dance hall. Soon, the fuselage and wing planes were rebuilt from spares the team had brought over with the aircraft. The Falcon engine was taken apart and subjected to a minute examination. To the amazement of the engineers, it was found not to have suffered significant damage in the crash. Though a replacement was on its way from England, the original was recondi-tioned and put back together.

There remained, however, the matter of finding a replacement navi-gator. Charles Morgan waved goodbye to Newfoundland on May 29 as he started his journey back to England for medical treatment. He arrived home on June 9, to be met by Phyllis and the children. Taking his place in St. John's would be a twenty-five-year-old sailor-turned-aviator by the name of Conrad Biddlecombe. Biddlecombe had already seen his fair share of dramatic action during his career. As a midshipman before the war, he survived a shipwreck in the Indian Ocean. After war broke out, he spent three years as a navigating officer in the Aegean and Mediter-ranean before transferring to the air service. In that role, he was shot down twice. Following the Armistice, he took on a junior position in the Air Ministry's navigation department. But he was eager for more adventure. When Martinsyde offered him the transatlantic navigator's job, he jumped at the chance, accepted the role, and got ready to head for Newfoundland to join Fred Raynham.

Over the days that had passed since the Martinsyde crash, people at St. John's observed an interesting change in Raynham's character, a change that took them by surprise. Until the fateful moment when the Martinsyde men attempted their take-off, twenty-five-year-old Raynham had come across as quiet and reserved. He wasn't impatient or a joker, like Harry Hawker. He didn't yet possess Mac Grieve's avuncular character. He certainly had none of Charles Morgan's garrulousness, a cheerful optimism that must have carried the Martinsyde team forward as they first prepared the aircraft. With Morgan keeping the mechanics motivated, Raynham hadn't seemed to have much to say. He appeared, perhaps, a little colourless. But after the crash, he had risen from his bed a new man. He seemed energized by his defeat. He was newly communicative. He had developed the ability to manage a team, something he had never done before. Now that Morgan was gone, Raynham, enlivened by the events that knocked him out of the race, had become a leader.

People who had written Martinsyde off after the crash, the ones who thought the only contest now was between Handley Page and Vickers, started to realize that Fred Raynham was still in the game. He just needed his new engine, and his new navigator, and he would be ready to go again.

THE SUCCESS OF ALCOCK AND BROWN'S TEST FLIGHT AND THEIR perfect landing at Lester's Field on Monday the 9th had lifted the Vickers men. But Mark Kerr was close behind. Handley Page had made good progress over the weekend. The weather had been kind to them: it was warm and bright, with a breeze from the south-west. Sunday afternoon crowds descended on Harbour Grace to watch the men work on the aeroplane. The fuselage had received its final coat of paint and was looking pristine. There was an air of festivity at the aerodrome, which spurred the mechanics on.

On Tuesday, Kerr decided that the big bomber was ready to make its first flight. Mechanics put a light load of fuel on board. The machine's four Rolls-Royce engines were started. Then Herbert Brackley climbed in, taking the pilot's seat. The rest of the crew followed. Brackley called for the chocks to be withdrawn, opened the engines to full throttle, and began the take-off run. A large crowd had gathered to watch the event. The giant machine took off in front of them after a run of just thirty yards, and climbed, said one reporter, "almost straight upwards." The onlookers cheered wildly as the plane left the ground. For fifteen minutes, Brackley circled over Harbour Grace to gain height. Reaching 2,000 feet, he then headed south-east for St. John's, arriving over the streets of the city fifteen minutes later. Nobody could escape the roar of the four Eagle engines. Once more, residents came out into the streets to watch the spectacle of an aeroplane passing overhead. When Brackley brought the aircraft back down onto the Harbour Grace airfield after an eighty-five-minute flight, Kerr's team was jubilant.

At Lester's Field, Jack Alcock and Ted Brown looked up, startled, as the aircraft flew over. They had been convinced that the Handley Page team needed a few more days to get their huge biplane ready. Now they realized the competition was closer than they thought.

Throughout Wednesday, the Vickers team laboured at their tasks, trying to solve myriad technical problems. During the afternoon, crowds enjoying a half-day holiday gathered once more at Lester's Field. The Vickers mechanics tried to ignore the distractions. At dusk, bright flares were stuck in the ground to provide illumination for the men. There was still so much to do.

The next day, however, the Vickers party caught a break. The ship carrying their replacement fuel arrived at noon. It also brought Percy Maxwell-Muller, manager of the transatlantic project and Alcock's old friend from before the war. They were all glad to see him. It had been a tough stretch since the aircraft arrived and they needed a boost. Brown

The Vickers Vimy On Lester's Field.   St. John's N.F. Copyright

Crowds around the Vickers Vimy at Lester's Field.

later described Maxwell-Muller as a "rabid optimist, with the power of infecting others with his hopefulness." He brought fresh energy when it was needed most.

The arriving ship also brought the replacement engine for Martinsyde. It didn't take long to cart it to the Blue Puttee dance hall, where Fred Raynham's mechanics could fit it to the rebuilt aircraft. They had given the machine a new name as well. With Charles Morgan off the scene, it could no longer be known as the Raymor. Instead, it was christened the Chimera. Whether it would breathe fire or prove illusory wasn't yet known. Conrad Biddlecombe was expected on the same ship that brought the engine. But he wasn't on board. Word was he might not arrive until Saturday evening.

On Thursday afternoon, Alcock and Brown made a second trial flight, lasting thirty-eight minutes. The take-off, the first from Lester's Field, went well. At one point during the flight, Alcock opened the engine throttles full. The machine bounded forward, reaching a speed of 120 miles per hour. Worried about the sediment that had blocked Harry Hawker's radiator and put an end to the Sopwith flight attempt, Alcock had carefully filtered and distilled his own coolant water before filling the Vimy's system. It had been a laborious and time-consuming task. But the faultless performance of the engines over Lester's Field proved its worth. There were no blockages.

After they landed, Alcock and Brown heard that bad weather over Harbour Grace had delayed Herbert Brackley's second test flight. The threat from Handley Page seemed to be receding, though there was always a risk they would abandon their second trial and make the trans-atlantic attempt without it.

The predictions from Lawrance Clements at the weather bureau that day were hopeful. Though he had scant information to work from, he considered that the winds over the Atlantic were likely to be in the favourable west-to-east direction for the next few days. This was all that Alcock and Brown wanted to hear. The aircraft was behaving perfectly. The weather looked as good as it was ever likely to get, and they were determined to beat the better-resourced Handley Page team. They resolved to depart the next day, Friday the 13th.

Lester's Field was seething with activity that day. The Vickers team worked on their final tasks to prepare the Vimy to fly. Everything was checked and adjusted. Finally, the aeroplane was ready to go.

The capricious St. John's weather chose this moment to put a stop to proceedings. Without warning, the wind, which had been coming from the perfect direction, suddenly changed course and began to blow across the take-off strip. With a fully laden aeroplane it would be risky to take off in a crosswind. Fred Raynham had shown the danger of such a manoeuvre. Alcock was exasperated, telling nearby reporters, "It's maddening." The airmen would have to wait for the wind to turn. Then Alcock and Maxwell-Muller spotted a problem in the right-hand under-carriage. Its axle was slightly twisted. Normally, such a fault wouldn't cause a problem. But with such a heavily loaded aircraft there could be no question of leaving it unrepaired. Scaffolding would need to be built underneath the fuselage and wings to allow the body to be lifted from the undercarriage. There was no other way to do it safely, which meant there was no way they would finish before the sun set that day.

WHILE FRIDAY PROVED FRUSTRATING FOR JACK ALCOCK, IT WAS worse for Mark Kerr. He and his colleagues at Harbour Grace finally decided to go up for their second test. A little after noon, they climbed into the cockpit and settled in for a five-hour endurance flight. After a 290-yard take-off run travelling uphill, the aircraft made it into the air without trouble. But soon, Brackley, who was piloting the flight, made a troubling observation. The entry he made in his logbook spelled out his concerns in spare prose: "Radiators gave trouble, had to throttle constantly to prevent boiling. Landed after 1½ hrs. flying to get rectified."

For the next four hours, mechanics laboured over the radiator assembly, stripping off lagging from the water tanks to try to reduce the overheating problem. Then the airmen returned to the cockpit and resumed the flight. "Slight improvement," Brackley logged, "engines boil if left on full throttle for longer than 10 mins." For almost five hours, the airmen drove the giant biplane through the skies above Newfoundland, nursing the engines to discover at what level of power they might function adequately. Kerr and Gran took turns at the controls to get a feel for the aircraft. Then they landed once more at Harbour Grace. Brackley noted his conclusion in the logbook: "These conditions not satisfactory for Atlantic Flight." It was, in Kerr's later words, "the worst blow that could have happened."

They had already known the radiators might give trouble. There had been a defect in their manufacture, and a replacement set had already been despatched from England, due to arrive at St. John's any day. But even a short delay risked putting Kerr's crew out of the contest. With Alcock and Brown almost ready to go, and with reports of good weather over the Atlantic, it looked as if Handley Page would be beaten to the post.

But Vice-Admiral Kerr had proved that he was not the type to yield without a fight. He gathered his team and asked what could be done to sort out the cooling problem using only the resources available at Harbour Grace. As they talked it through, the mechanics started to form plans for a makeshift fix that might keep the engines just cool enough for the transatlantic flight. Though it offered the possibility of success, the fix would take at least a day to carry out. Nobody wanted to back out now. They all agreed to proceed with the work. They also set a deadline. The aircraft would need to be ready to go at 5 p.m. the following day, Saturday the 14th. It was the same day that Jack Alcock intended to get away. The competition was heading for the wire.

CHAPTER 17

A HURRIED LINE BEFORE I START

JACK ALCOCK AND TED BROWN LEFT LESTER'S FIELD LATE ON THE afternoon of Friday, June 13, and went back to the Cochrane Hotel. Neither of them had managed to get a decent night's sleep for weeks. They worked late into the evenings most days, and the last fortnight had been particularly demanding. They needed an early night. They felt bad about leaving the mechanics working at the aerodrome, because they knew the men had a long stretch ahead of them. But there was no point doing any of it if the pilot and navigator weren't on the top of their game. At 7 p.m., the airmen asked hotel staff to wake them before dawn, then turned in.

They slept soundly for eight hours. At 3 a.m., they rose, packed the few items they planned to take with them on the transatlantic flight, and dressed. Alcock wore a blue lounge suit with a soft tweed cap. Brown had chosen to wear his horizon-blue Royal Air Force uniform. Then they got into the old Buick and headed for Lester's Field for what they hoped would be the last time. They got to the aerodrome in fine spirits.

Strong, gusty winds were coming from the west, too strong to allow an early departure. As he waited for the wind to lessen, Alcock chatted with the mechanics who had worked through the night. Dawn broke at about 5 a.m. As the watery Newfoundland sunlight began to creep over Lester's Field, Alcock inspected the work that had been carried out. He spotted a problem with the right-hand axle that had been replaced the previous day. An elastic shock-absorber wasn't working as it should, and it had to be changed. Later, after the job had been completed to Alcock's satisfaction, the aircraft had to be moved across the aerodrome to a new starting position dictated by the wind direction. As the engineers and mechanics pushed and dragged the huge machine across the uneven ground, a rope became fouled around one of the undercarriages, crushing a copper fuel pipe. The buckled section was cut out, then one mechanic held his hand over the open end of the pipe to hold back the petrol while others jointed in a replacement section. It took an hour for the repair to be made. Once more the departure was delayed.

But these setbacks did nothing to upset the moods of the two airmen. They sat on the grass nearby, chatting with members of their team and onlookers. They were relaxed and cheerful. At one point, Alcock pulled a handful of grass from the turf and threw the blades into the air, watching as they were carried away by the gusty wind. He turned to the waiting journalists and explained that it was a pre-war technique. "We would toss a handful of torn tissue paper into the air," he said, "and if there was the slightest puff of wind we didn't go up. It is different now; I'm going."

The airmen also chatted with photographers who had been documenting the contest since the Sopwith party first arrived back in March. Among them was twenty-three-year-old Margaret Carter, granddaughter of a former Newfoundland prime minister. Carter had grown up in a house just off Duckworth Street, a few paces from where Lawrance Clements would later set up his weather bureau. She took an intense interest in aviation. With access to a motor car and a camera, she spent the spring and summer of 1919 travelling endlessly between Mount Pearl, Pleasantville, Harbour Grace, Lester's Field, and the Cochrane Hotel. Over the weeks, she came to know the aviators well. They dined together, talked together, and no doubt shared their hopes and fears together. Also present at Lester's Field was thirty-seven-year-old Elsie Holloway, a professional photographer with a studio on Bates Hill, just moments from Duckworth Street. Holloway had made an extensive record of the Vickers team as they laboured over the Vimy's assembly and its test flights.

She, too, had got to know Alcock and Brown well. Her photographs captured unguarded, informal moments, revealing character that would be lost in formal portraiture.

After a few hours, repair work on the Vimy was completed. The wind, though still strong and gusting, was slackening. It was time to get ready. As Carter and Holloway moved around the aircraft, recording the activity with their cameras, Alcock and Brown loaded provisions into the cockpit. They took flasks of coffee, hot chocolate, and Horlick's. In their pockets they carried sandwiches that Carter had wrapped for them in a linen cloth, as well as water and a small flask of brandy each. In a special cupboard at the tail of the aircraft, they packed further food and drink in case the aircraft had to come down on water. They loaded the sealed mailbag into a cupboard at the back of the cockpit. Among the letters it contained were several written by the two airmen on Cochrane Hotel notepaper. Brown had penned one to his fiancée, Kathleen, and Alcock had written to family members. To his younger sister, he wrote, "My Dear Elsie. Just a hurried line before I start. This letter will travel with me in the official mail bag, the first mail to be carried over the Atlantic. Love to all. Your Loving Brother, Jack." He wrote to his mother asking her not to worry.

Next came a task that was small but important to the two airmen. Each put a mascot on board. Alcock's was a black-cat toy with a big head, which he named Lucky Jim. He knotted a piece of red, white, and blue bunting around the toy's neck before stowing it carefully in the tail cupboard. Brown brought a black-cat mascot as well. It was from Kathleen, and he called it Twinkletoe. Like most airman veterans of the war, Alcock and Brown were sincere in their superstitious belief that black cats brought good luck. One American airman recalled the mascot he carried during aerial battles. He wrote, "It always gave me a lot of courage in a tight place to see the very placid, untroubled expression on her or his face. If the cat wasn't worried, why should I be?" As well as Lucky Jim and Twinkletoe, Alcock and Brown each carried a tiny black-cat charm on their person.

They packed other charms and keepsakes given to them by friends

Jack Alcock beside the Vimy, photographed by Elsie Holloway.

and loved ones. In one of his pockets, Alcock carried a little silver kewpie doll. He also hung two little yarn dolls in the cockpit, one named Ran-Tan-Tan, the other called Olivette. Brown carried a tiny cardboard aeroplane made by one of the boys working as a clerk at the Ministry of Munitions. The boy had told Brown he wanted an aeroplane he had made to fly the Atlantic. The men also stowed bunches of white heather they had been given by well-wishers. Whereas Alcock tied British colours around Lucky Jim's neck, Brown, an American citizen, carried a little stars-and-stripes flag. Kathleen had given it to him, along with Twin-kletoe, before he departed for Newfoundland. His last words to her had been, "I have a hunch that I'm going to win." He held her tiny silk flag in his wallet, close to his heart.

As activity on the airfield drew to its final stages, Alcock and Brown stopped to eat a meal before departing. They sat down on the grass under the wings of the Vimy and ate cheese, bread, cold tongue, potted meat, and crackers, with hot coffee to follow. They expected it to be their last meal in Newfoundland. The wind was still blowing hard, but after waiting so long, Alcock and Brown resolved to go anyway.

After finishing their meal, the pair stood up and started to say their goodbyes. There were not many spectators at Lester's Field that day. The high wind had convinced St. John's residents that the departure would once again be delayed. But their friends had come to see them leave. Fred Raynham was there to see Jack, his old friend, off safely. Percy Maxwell-Muller and Bob Dicker looked on warmly. Lawrance Clements, the Royal Air Force meteorologist, was there to say goodbye, as were several other specialists the pair had encountered during their sojourn on the island. Also waiting to see the airmen off were five of Newfoundland's top politicians, including Prime Minister Michael Cashin and the mayor of St. John's, William Gosling.

Then Alcock and Brown turned to the mechanics and engineers who had worked so persistently and without complaint over difficult weeks. From the start, Alcock had inspired the men, and they had inspired him in return. Alcock was the sort of person who led from the front. He was decisive. He got involved with every job. He worked the same long hours he expected of his men. And he remained unflappable, even when things went wrong. He had no affectations, and he wore his skills and experience modestly. Brown was held in high esteem by the men as well. He was more highly strung than Alcock, everyone had seen that, but he kept his anxiety under control, and he was amiable and cheerful. He would help with any job that needed attention, however small or mundane. And he prepared earnestly and studiously for the navigation task ahead. The mechanics appreciated it all. Earlier that day, one of them nailed a discarded horseshoe found lying in Lester's Field to the underside of the Vimy's seat. It was a final quiet gesture of hope and goodwill.

Finally, Alcock addressed the waiting reporters. "It's a long flight, but it doesn't worry me," he told them. Then the pair said their closing words: "So long."

They got into their flying clothes. Over his lounge suit, Alcock put on a waistcoat and gloves that would be heated by a battery in the Vimy's cockpit. Inside his fleece-lined knee-length flying boots were heated insoles. He also wore a life-vest. Then he pulled on his windproof flying suit. It was a dull grey-brown colour, made by Burberry, loose-fitting with numerous pockets. Brown wore the same garments as Alcock. The only difference was the air force uniform he wore under them. Brown had to be helped into his flying clothes by Bob Dicker. He had suffered an injury to his foot during the war which limited his mobility.

Alcock looked out at the spectators lining the field one last time. Among them, he spotted a local boy, brought by his parents to witness the event. The youth had one arm stretched high in the air. In his hand was a model aeroplane, its propellers spinning fast in the wind. Alcock called him over. The boy told him he had made the model himself. It was a Vimy. Alcock took it into his big hands, then turned it round and around, examining every part intently. His ruddy face had broken into a big grin. Perhaps he recalled his own boyhood, in Manchester, when he made model aeroplanes and balloons and watched them fly. He handed the little Vimy back to its owner and thanked him.

With that, it was time to go. As the wind continued to blow down the airfield, Alcock told onlookers, "Oh, well, if we crash we won't be going fast against that gale, anyway." Brown just smiled. Then the pair put on their fur-lined flying helmets, climbed up into the cockpit, and took their seats, side by side. Alcock looked down and saw Bob Dicker standing nearby. He shouted, "Bob, what are you doing here? Get them started." Dicker called for the port engine to be run. It took a few attempts to get it going, but soon it was firing away happily. Dicker let it run for a while as Alcock and Brown made their final checks in the cockpit. Lawrance Clements passed Brown the latest weather charts. They predicted a lessening westerly wind and clear weather over most of the Atlantic. After twelve minutes, Dicker gave the instruction to start the starboard engine. It sparked into life right away. For six minutes, as Alcock ran the engines up and down to warm them, the Vimy sat motionless on the grass, its wheels chocked, mechanics hanging on to the wings and the tail.

Finally, Alcock gave the signal that he was ready to depart.

The chocks were pulled away, the mechanics let go of the aeroplane, and Alcock opened the throttles to full. The aeroplane began to roll across the bumpy field. There were fences on either side of the take-off run and at the end were trees, more fences, a low wall, and buildings. The wind was still gusty and unpredictable. Occasional flurries would come from the sides, trying to tip the aircraft or push it off course.

As the machine lumbered along, gradually accelerating, Alcock could be seen looking firmly ahead, his face fixed in concentration. The onlookers stared nervously as the Vimy bumped and laboured across the grass.

The halfway point was reached, yet the machine was still on the ground. It was running faster now, its momentum as well as the thrust from its powerful engines carrying it forward. The end of the field was getting closer as the aircraft continued to accelerate. The spectators watched as it approached the low wall and the fence and the trees at the end of the course. Their expressions started turning from anxiety to alarm. Everyone looked intently at the four heavy wheels as they ran over the ground. The end of the aerodrome was almost upon them. The low wall was just fifty feet away and closing fast. Suddenly, slivers of daylight could be seen separating the wheels from the grass. To cheers and cries of relief, the Vimy had become airborne.

But it only hopped five feet off the ground. It dropped down, hit the ground with a bump, then bounced back up. It cleared the wall by a mere two feet. Buildings and trees were dead ahead. Then the aircraft fell and disappeared behind the crest of the hill.

A cry went up from dismayed spectators: "They're down!" There was nothing to do but watch and hope. The Vimy appeared once more, rising slowly from the hollow into which it had dipped.

In the cockpit, Ted Brown stole a sideways look at Jack Alcock's face. It was running with sweat.

Alcock was struggling to gain height. As the heavy aircraft headed down the valley towards the neighbourhood of Topsail, due west of St. John's on the Conception Bay coast, it dipped and rose several more times, each time disappearing and reappearing as the spectators at Lester's Field looked on. The gusty wind was rocking and buffeting the craft as it strained to rise.

By now, the machine had reached Topsail and was headed west towards the bay. Because he was flying so close to the ground, Alcock couldn't risk turning the aircraft back in the direction of St. John's and the ocean. He kept flying straight ahead, hoping for better conditions when he emerged from the valley. Over Portugal Cove, the aeroplane disappeared from view once more—and didn't reappear.

At Lester's Field, Bob Dicker began running towards the Buick. He shouted to Percy Maxwell-Muller, "I'm going to find them." Fred Raynham ran off with him. The pair piled into the vehicle and drove off towards Topsail Road. Some of the spectators followed them. As the convoy of cars raced towards Topsail, those left behind at Lester's Field searched the sky with a growing dread. The minutes ticked by, each one an eternity. Dicker and the other motorists peered into the scrubby brush as they drove, hunting for signs of the aircraft. The wind was starting to howl through the valley.

Back at the aerodrome, the mood was hollow. Eleven minutes had passed since the aircraft had last been seen.

On Topsail Road, Bob Dicker and Fred Raynham stopped and got out of the Buick. They were peering into the sky in search of the Vimy. Then Raynham exclaimed, "Look Bob." Through a gap between the hills lining the valley, they saw the aircraft flying east at about 800 feet, gradually gaining height as it approached St. John's. Somehow, out over the bay and out of sight of the frantic onlookers, Alcock had managed to turn the aeroplane around and gain height. Dicker recalled, "It looked bloody marvellous."

With the wind now behind it, the Vimy was barrelling along at a speed approaching 130 miles per hour. It acted as if it was a little fighter scout, not a heavy two-seat bomber. It passed over Lester's Field. As Alcock concentrated at the controls, Brown looked down, waving hard. A roar of unchecked relief came from everyone gathered there.

The Vimy flew on. It passed over the city to the harbour, now high in the sky. It crossed the harbour mouth at 1,200 feet and reached the open ocean. The time was 4.28 p.m., Greenwich Mean Time, the timescale that would be used to log events throughout the journey. To the watchers in

The start of the flight, June 14, 1919.

St. John's, the aeroplane dwindled until it was a silver speck in the sky. Then it was gone.

In the cockpit, Brown looked once more at his chart. He was aiming for Galway Bay, halfway down the Irish coast. Before leaving the land behind, he had wound out the wireless antenna using a crank in front of him and keyed his first Morse message into the transmitter: "All well and started."

PART THREE

We don't yet know how far this new science will alter the world, it must rub boundaries off the earth, nations cannot help getting mixed, and what effect it will have on laws, character and trade are mere speculations.

—HILDA HEWLETT, AVIATOR

CHAPTER 18

WE MUST GET STARS

AT THE START OF THE FLIGHT, THE OCEAN RESEMBLED AN INDIGO carpet, dotted by glistening white icebergs and lit by flashes of sunlight breaking through the clouds above. It soon began to take on a hazy purple hue as the aeroplane passed over it. Before long, any trace of ocean blue was hidden altogether. Above the aircraft now was thick cloud, getting thicker with each passing minute. The sunlight which gleamed on the silver-doped wings and taut steel wires a few minutes earlier had disappeared. The world that existed around the Vimy was now a flat mass of dull grey vapour. The aeroplane had entered what Jack Alcock once described as "the dreaded fog of Newfoundland."

Ted Brown spent the first forty-five minutes of the flight taking advantage of the good visibility by getting as many navigational fixes as he could. The sky gave him astronomical data. To measure the height of the sun and stars above the horizon, he used a marine sextant with a scale engraved more deeply than usual. This allowed him to read it better in the cramped conditions of an open cockpit travelling into the night. When he couldn't make out the horizon distinctly enough, he used a spirit level. Though less accurate than the real thing, it gave a good approximation. Looking down to the sea, Brown used his drift indicator to measure the direction in which whitecaps or icebergs appeared to be moving as the Vimy passed overhead. From this, he calculated how the wind was causing the aeroplane to drift off course, so he could work out a correction. With his stopwatch, he also used these drift measurements to compute their ground speed. Brown made most of these observations by kneeling on his seat and looking back between the aircraft's wings or over the side of the fuselage to the water below. Then he squeezed back down onto the seat next to Alcock, positioned his chart and navigational calculator on his lap, and worked out their location.

But when they entered the fog bank, he was cut off from sky and sea and the navigational information they gave him. As Alcock piloted the heavy machine onward, Brown concentrated on his computations. Until the clouds or the fog parted, he would have to navigate by dead reckoning—a process of well-educated guesswork involving estimates of wind speed and direction.

It was more important than ever that Brown be able to communicate with the outside world by wireless. Before leaving St. John's, he arranged to radio the aircraft's position at twenty minutes past each hour. That way, the airmen could be tracked along their journey. If they had to come down onto the sea, help would not be long in finding them.

From the moment the Vimy crossed the Newfoundland coastline and headed out over the wide ocean, operators at two wireless stations, one at St. John's and a second down the coast at Cape Race, had been listening intently to their receivers, waiting to pick up the Vimy's hourly transmissions. Out in the ocean, three steamships lay along the aircraft's intended route: *Digby* was 100 miles out when the airmen set off; the cable-steamer *Mackay-Bennett* was working 250 miles offshore; and the steamship *Sachem* was 700 miles from the Newfoundland coast. Each was listening out for Brown's transmissions as well. But his messages never arrived. Each passing hour just brought silence.

At 5.20 p.m., as he noted in the navigator's logbook, Brown had started tapping his first hourly message into the transmitter's Morse key. But just a few taps in, the machine suddenly stopped working. At first, Brown was unsure what had happened. He searched for clues in the cockpit installation. Then, thinking the problem might be in the windmill-powered generator clamped to one of the aeroplane's struts, Brown looked round to check that its little propeller was still spinning. To his dismay, he saw that it had sheared off and fallen away. The transmitter had no power. The airmen could no longer send messages. They still had a battery-powered receiver, but now, just an hour into their journey, they had been struck dumb, just as Harry Hawker and Mac Grieve had been. Whatever happened now, whatever challenges they faced, nobody would hear them. Jack Alcock and Ted Brown were entirely alone.

They cruised on through the fog. Alcock kept his hands clamped on the wheel at the top of the control column. This assembly operated the hinged ailerons on the wings for banking and the elevators at the tail for climbing and descending. His feet controlled the tail's rudders through a bar set into the cockpit floor. This turned the aeroplane. A lever at his right hand operated the engine throttles and fuel mixture controls. A crank on the side wall of the cockpit adjusted the radiator shutters, giving him control over the engine temperature. In front of him, the control panel was sparsely populated in modern terms. There were two compasses and an air-speed indicator. An air-pressure meter known as an aneroid gave Alcock an approximation of the aircraft's altitude, though it couldn't be relied on to be totally accurate. There was a water temperature gauge for each engine. Then there were controls and gauges for the ignition and fuel systems.

Outside the cockpit, fitted to the fuel line passing down one of the central struts, was a petrol overflow glass. This gave a visual indication that fuel was passing safely from a header tank down to the engines, and it was a vital safety feature. Installed in the sides of the engine nacelles were oil pressure and engine speed indicators. At Brown's side of the cockpit were the wireless sets, antenna reel, and battery.

There was little legroom and certainly no space to move about. Though Brown could kneel uncomfortably on the tiny seat when taking readings, Alcock was completely stuck. An elastic device had been fitted to the controls. Its purpose was to keep them in the optimum position during much of the flight, relieving the pressure on Alcock's arms and legs. But in the race to complete the aircraft assembly at St. John's, the elastic had been cut too short; the device was useless. This meant that Alcock would have to bear the full weight of the control surfaces throughout the journey. Moreover, as the fuel burned and the tanks behind the airmen emptied, the trim of the aircraft would shift, causing the centre of gravity to move forward. The aeroplane would want to nose downward; Alcock would need to pull back ever harder on the stick. It was going to be a long, uncomfortable flight.

The fog was thick and getting thicker. There was nothing to see beyond the aircraft. The sky and the sea were hidden. If they had not had the drone of the engines by their heads, and the needle of the air-speed indicator showing their forward motion, the two men might as well have been hanging motionless in the grey, indistinct space. Without his calculations, Brown would have had nothing to occupy his mind. Without the constant pressure of the control column and rudder bar on his limbs, Alcock might have forgotten he was piloting an aircraft at all. The first hour and a half passed like a dream.

Then, without warning, the two men heard what sounded like machine-gun fire. It was coming from a point directly beside Alcock's right ear. A staccato clattering sound filled the air, getting louder and louder. Within seconds it was deafening. The two men, alarmed, looked out at the starboard engine. The exhaust pipe had burst. A large section of it had broken away and was hanging by a fragment. It was glowing red with the heat from the exhausting gases. Then it turned white-hot. The noise became jerkier as the three engine cylinders that exhausted into that section of pipe were now open directly to the air. As Alcock and Brown watched, the exhaust pipe softened in the intense temperature, crumpled, and blew away in the slipstream. The noise now swelled into a roar. Alcock forced his attention back to the controls, evaluating the consequences of the burst exhaust in his mind. Brown kept looking behind Alcock's head at the damaged engine. He saw fire emerging from the open exhaust pipe. The flame was playing directly onto one of the tensioning wires that braced the aircraft together. The wire was starting to glow red from the heat.

———

ONCE THE VIMY PASSED OUT OF SIGHT, BOB DICKER AND PERCY Maxwell-Muller went off with Lawrance Clements to the wireless facility at Mount Pearl. Together, they began sending wireless messages after the aeroplane, then listening intently to the receiving equipment for a reply, but no response came. The men suspected that the radio silence was down to some kind of fault in the Vimy's wireless equipment. They knew the devices were temperamental and prone to failure for numerous reasons. But it was hard to keep their minds clear of more worrying explanations for the absence of messages as time went on. The plight of Hawker and Grieve's aircraft still occupied everybody's thoughts.

After a few hours, Dicker and Maxwell-Muller left Clements at Mount Pearl and returned to Lester's Field, where the Vickers mechanics were clearing up. They were not sure what to tell the men. This would be their first free evening in three weeks, and they badly needed a break. Dicker and Maxwell-Muller knew that nothing they said could change the course of events playing out over the ocean, whatever it was. So, they resolved to hide the truth behind a white lie. They said that no message had been received from the aeroplane directly. Then they told the mechanics they had picked up a transmission from a ship 500 miles out at sea that reported hearing an aircraft fly overhead. The mechanics, cheered by this fiction, went on to enjoy their evening. Dicker and Maxwell-Muller knew there was nothing anybody could do but wait.

IN THE VIMY, ALCOCK AND BROWN WERE SETTLING INTO THE NEW reality of their flight. Until the exhaust fractured and fell away, the airmen had been communicating with each other using an intercom system fitted to their flying kit. Earpieces inside their fur-lined helmets were shielded from the expected noise of the engines. Microphones strapped to their necks picked up speech directly through throat vibrations. Alcock never liked using the system. Anything tight around his neck made him uncomfortable. The damage to the exhaust pipe and the din from the engine made the intercom even harder to get on with. After a while, the pair fell back on gestures, for the most part. Sometimes, Brown would make short, pencilled notes on his pad and show them to Alcock. There wasn't much to say in any case. Each man had certain knowledge that the other knew what he had to do and could be left to get on with it. They trusted each other absolutely.

By about 7 p.m., two and a half hours into the flight and an hour after the engine exhaust failed, Brown had grown concerned about the aircraft's navigation. He asked Alcock to climb through the layer of clouds hanging above them at 2,000 feet. He hoped they would find clear sky above. Alcock obliged, gently nursing the Vimy upward through the cold, white vapour. Moisture condensed on the instrument dials and the bracing wires. He kept the engines throttled back to no more than three-quarters of their full speed. The tail-wind that had hastened them on their way since Newfoundland meant there was no need to strain the engines; they were making good time and the Vimy seemed happy. After a little while climbing through the cloud, the aircraft emerged from its top layer at about 3,000 feet. Brown looked upward with the eager anticipation of seeing the sun above them once more. Instead, there was another layer of cloud. Alcock estimated that it was at 5,000 feet. He levelled the aircraft off, and they flew onward, now sandwiched between a white ceiling and a white floor.

Neither man had eaten or drunk anything since their airfield picnic back at St. John's. Brown now reached into the cockpit cupboard behind him and got out some sandwiches and chocolate, a serving of which he put into Alcock's left hand. After the pair had eaten, Brown opened one of the vacuum flasks so that they could drink hot coffee.

It was a quick meal. Once they finished, Brown put the flask and food wrappings away, then passed a note to Alcock which read, "If you get above clouds we will get a good fix to-night, and hope for clear weather to-morrow. Not at any risky expense to engines though. We have four hours yet to climb."

Alcock took him at his word and climbed only slowly. At 8.30 p.m., still flying in the gap between the two cloud layers, those above them suddenly parted. The sun came streaming through, casting a shadow of the Vimy onto the lower clouds. The shadow twisted and contorted with the bumps and hollows of the cloud surface. It was a much-needed reminder that they were making progress. Brown picked up his sextant, turned round, and knelt on the seat. For the next ten minutes, he sighted the sun through the gap between the port wings. Then he sat back down and set to work making calculations.

Just then, as the gap in the clouds above the Vimy closed, the clouds below them parted for a few moments. Brown grabbed his drift indicator, poked his head back over the side of the fuselage, and took readings of the sea. He worked out their drift and ground speed, fed the figures into his calculations, and computed their position. He discovered that they were flying farther east and south than he had calculated. His dead reckoning had been badly off. He had assumed that the tail-wind would lessen through the journey, as Lawrance Clements had predicted before they departed. Now he realized that Clements had been wrong. The wind had strengthened, not weakened.

At 9.30 p.m., with the daylight starting to dim, Brown asked Alcock to make the climb through the upper layer of cloud, which he estimated would take them to 6,000 feet. He added, "We must get stars as soon as poss."

Alcock began to push the aircraft upward. With the daylight fading rapidly, the clouds, already dense, turned darker and more ominous. Their structure was no longer discernible. As the aeroplane rose through the gap between the cloud layers, it became lost in a mass of vapour.

They kept climbing. The last remnants of daylight winked out and the world around the Vimy turned to darkness, lit only by the faintest glimmer of the moon. All the men could see outside the cockpit were the engines, their bright-red exhaust gases, the struts and wires closest to the fuselage, and the nearest parts of the wings. At midnight, they were at 6,000 feet. There was no change to the blackness. Brown had never felt more isolated.

Just after midnight, the Vimy emerged from the upper layer of cloud. Once again, Brown looked upward, hoping to see a clear sky. Instead, there was a third layer of cloud, this one several thousand feet higher. But it was patchier than the lower layers. Within minutes, the clouds cleared into a gap, revealing a bright carpet of stars. Below the Vimy, the clouds through which they had just climbed spread out flat and horizontal, lit brightly by the moon. They gave Brown a perfect cloud horizon. He took up his instruments, knelt on the seat, and aimed his sextant first at Vega, then at the Pole Star. He computed their position, compared it with their intended course, and passed a course-correction instruction to Alcock. On the note, he told his companion excitedly that they were halfway.

Now that the navigation was back on track, Alcock was able to take the strain off the engines and gradually drop down to a lower altitude. He took it slowly. After an hour, they had reached 4,000 feet. An hour after that, they were at 3,600 feet. The clouds above were patchier, though the aircraft was once again passing through fog. Occasionally, shafts of soft moonlight would break through, bathing the Vimy's wingtips in silver, gold, and red. The shadow of the aircraft on the lower cloud layer danced and writhed.

It was now the early hours of the morning, about 2.30. Neither airman was tired. Though the air outside was cold and damp, Alcock and Brown were snug inside their thick flying gear. The engine radiators offered warmth and the windscreen sheltered them from the oncoming blast. They were uncomfortable from sitting in such confined space for so many hours. It was bad enough for Brown, whose wartime injury troubled him. Alcock couldn't even move his limbs; they had to stay planted on the controls. However, the discomfort was bearable. In a quiet moment, Brown reflected on his feelings. He wrote, "An aura of unreality seemed to surround us as we flew onward towards the dawn and Ireland. The fantastic surroundings impinged on my alert consciousness as something extravagantly abnormal—the distorted ball of a moon, the weird half-light, the monstrous cloud-shapes, the fog below and around us, the misty indefiniteness of space, the changeless drone, drone, drone of the motors."

He reached back to the cupboard containing their provisions and brought out more sandwiches and chocolate, followed by hot drinks. His mind cast itself forward to the sun that was soon to rise and the new day that would soon dawn. They had travelled for ten hours since leaving Newfoundland. The starboard engine had got no worse since the exhaust pipe broke away. It was noisy, but they had become accustomed to the roar. The bracing wire heated by the exhaust flame seemed to be fine. There was plenty of petrol in the tanks, and Alcock had been cosseting the engines all the way, keeping them throttled well back and comfortable. They had taken good sightings of the stars, and they were on course. Halfway through the flight, all seemed well. They only hoped nothing unforeseen would shake them from their steady path.

At ten minutes past three, something did.

THEY WERE FLYING AT 3,500 FEET WHEN, WITHOUT WARNING, they ran into a thick bank of fog projecting upward from the cloud layer below. In the darkness, they hadn't seen it coming. Suddenly, they could no longer discern the aircraft's wings or even the nose of the fuselage in front of them. They were flying totally blind. The wind behind them, getting stronger all the way, had been buffeting them for hours, causing the aeroplane to roll and pitch. Alcock adjusted the controls instinctively. But now he had lost the visual cues that enabled him to do so. The turbulence was rocking the aircraft from side to side. There was no horizon, no sky, no sea. Nothing but the oil blackness of the fog that suddenly enveloped them. Alcock glanced at the instruments mounted on the dashboard in front of him. Two of them were critical in moments like this. The aneroid gave a measure of the aircraft's altitude; the air-speed indicator provided a crucial external reference frame. Together, they would help him decipher what was happening to the aircraft's position as it bounced its way through the fog.

Then, just as he was looking at the needle of the air-speed indicator, it froze in front of his eyes. He could smell smoke. Its sensor, mounted above his head, had become packed with sleet and jammed. The indicator was now useless. The turbulent wind made the aircraft sway and judder. Alcock was fast losing his bearings. His instinctive understanding of the Vimy's position was gone. To try to get his equilibrium back, he drew back the control column, hoping to pull the nose up. The aeroplane hung motionless for a second. Then it fell into a steep spiral dive.

Alcock could see the compass needles spinning around rapidly as the Vimy tumbled downward. He could feel himself being pressed back into his seat. The roar from the engines became louder as their rotation increased to twice normal speed. They began to send vibrations right through the aircraft's structure. Quickly, Alcock reached for the controls and throttled both engines back. A glance at the aneroid showed that the aircraft was falling rapidly, but Alcock had no idea what course it was taking, nor at what angle. He tried to centralize the controls, but without external cues he had no reference to centralize against. Brown, frantic, searched the sky for a glimpse of something—anything—that might anchor them to their surroundings once more.

The aneroid kept falling. It reached 3,000 feet, then 2,000, then 1,000. The needle flickered at 500 feet. The bank of cloud might extend all the way down to the water. If it did, they would hit before they could do anything about it.

As they spiralled down, Brown loosened his safety belt and gathered up his notes. He and Alcock looked once more at the aneroid, and then at one another. It showed 100 feet—and it was still falling. Brown knew only too well what was coming, because it had happened to him before.

CHAPTER 19

A BAD LANDING

THE DATE WAS NOVEMBER 10, 1915. AS HIS AIRCRAFT DIVED, TED Brown could do nothing but sit, grim-faced and helpless, as the pilot, twenty-two-year-old Harold Medlicott, struggled with the controls. The thick clouds seemed endless, and strong winds kept buffeting the aeroplane as it descended. A heavy snowstorm had now left the pair freezing cold, sitting exposed in the open seats of their biplane in misery. Eventually, the aircraft emerged underneath the thick cover of clouds that had left them disoriented as they first began to fall. The engine that had begun to misfire now cut out completely. After what seemed like eternity plummeting downward, the aircraft finally crashed.

Brown and Medlicott had been carrying out a routine morning reconnaissance mission for the Royal Flying Corps over German lines in north-eastern France. It was impossible weather for flying. They shouldn't have gone out in it, but the orders from wing headquarters had been urgent and unequivocal. Brown had developed a reputation as a brilliant and gutsy observer, and Medlicott was simply the best pilot around. With misgivings, they had set off. But rain, thick cloud, and 60-mile-per-hour wind forced them to fly low. They were so low their B.E.2c biplane came within range of enemy shell and rifle fire.

Brown later recalled, "By the time we reached Vitry—where our chief observations were to be made—we got it hot and were heavily and con-

tinuously shelled." It was a discomfiting experience. Nonetheless, the
pair were determined to continue their mission, even though their two
protective escorts had disappeared some distance back. There must have
been a good reason why the supporting pilots disengaged from the mission. Perhaps they had suffered engine problems, or maybe they had been
shot down. It turned out later that they had got lost in the thick clouds
and had to struggle back to base. The reason hardly mattered. Brown and
Medlicott were now alone and under fire. They could only hope for the
best as Brown carried out his observations.

Once their task over Vitry was completed, the pair turned back
for home. Medlicott opened the engine's throttle full, racing the
aeroplane at top speed to get clear of the artillery barrage that now
surrounded them.

Royal Flying Corps pilots developed a range of manoeuvres they
could deploy when they came under the shellfire colloquially termed
"Archie." One aviator described the procedure they used to evade the
shells. "Automatically," the pilot wrote, "as the explosions cause his
machine to rock violently, he puts down the nose of his machine, banks
her over steeply, and spirals down some hundreds of feet to put 'Archie'
off the track. As the shells descend to his level, and commence to burst
around him, he puts his nose down slightly and rushes along at ter-
rific speed, suddenly zooming up almost perpendicularly and darting
away in quick, deceptive zig-zags, now up, now down, right and left,
successfully avoiding those murderous bursts from which hurtles the
quick death."

Harold Medlicott pressed on, weaving the little machine towards
the safety of home territory. For Brown, there was little to do except
wait, carefully protecting the sheaf of observation notes he had taken
during the reconnaissance. Shells from the ground came ever closer—
"puffs of whistling death" was how one pilot described them. Then a
shell exploded directly alongside Medlicott and Brown's aircraft. Hot
metal shrapnel peppered the machine as it rocked and shuddered from
the blast. One shard punctured the aeroplane's main petrol tank and the
precious fuel began to pour out. This caused the engine first to misfire

and then to fail entirely. Brown and Medlicott were now under sustained attack from the German gunners below. They found themselves amid a maelstrom of shrapnel and bullets.

Suddenly, Brown felt a searing agony in his left foot. He had been shot. This injury mattered little to him in the heat of the moment. If the pair were to survive this attack, the aeroplane's pilot would need to keep control of the broken machine. Two lives now depended on the skill and fortune of young Harold Medlicott.

Brown glanced back at Medlicott, fearful of what he might see. Once, a Royal Flying Corps observer at the front of an aeroplane under intense shellfire had grown alarmed by the straight, unchanging course his pilot was following. He'd expected him to follow the usual twisting-and-turning manoeuvres to avoid being hit. On that occasion, the observer looked back to see the pilot hanging dead in his seat, blood pouring from a shrapnel wound in his head. The inherent stability of aircraft like the B.E.2c meant they would fly straight and level by themselves until they ran out of fuel. That day, the observer had been forced to climb back along the fuselage into the pilot's seat. Then, while sitting in the lap of his comrade, he had to wrest the controls from the dead pilot's clenched hands and bring the aircraft in to land himself.

Brown was spared this horror as he turned to look at Medlicott. "Fortunately," Brown recalled, "he had escaped the shower of bullets which had broken round us like a hailstorm."

Their situation was grave. Without engine power, and with the aircraft now lost in clouds, Medlicott urgently signalled to Brown that they would have to make a forced descent.

The crash, when it came, was worse than either man had experienced before. "We made a bad landing," Brown said, "and I was thrown out on one of the crumpled wings." He was seriously wounded. "I had my leg broken, both knee and thigh dislocated, an injury to the lower bowel, lower lip cut, and several teeth knocked out," he reported. Despite the extent of his injuries, he managed to destroy the observation notes he had made over Vitry. At least they wouldn't fall into the hands of the enemy.

He looked over to see the fate of his pilot. He later observed, "Medlicott was hanging in a helpless position, head downward from his safety belt, which was fitted with a pin, and had no knife by means of which one could cut oneself adrift." Though unhurt, Medlicott was trapped, and Brown couldn't get to him.

It was just five weeks since Brown had been shot down over Vaudricourt. That time, he had been lucky to walk away. This time, he wasn't so fortunate. Before he and Medlicott could escape from the wreckage of their mangled biplane, they came under gunfire from the German soldiers who quickly surrounded them. They were forced to surrender.

Brown soon passed out from the pain of his injuries. "I next woke up to the fact that I was at a field dressing station," he recalled. There, his knee and hip dislocations were treated in a procedure known as reduction. The bullet that had entered his left foot underneath the heel, exiting from the upper surface of the foot, caused him agony. Medlicott was gone.

Brown was in a miserable state. He was lying in pain in a cold and muddy French field. He had been separated from the friend who had tried in vain to save them both. He was alone, broken, and captured.

FOR MUCH OF THE FOLLOWING NINE MONTHS, WHILE HAROLD Medlicott was incarcerated in German prisoner-of-war camps, Ted Brown received treatment for his injuries at a series of military hospitals.

Later, he was tight-lipped about his experience following the crash. He merely commented, "I knew the deadly heart-sickness which comes to all prisoners of war during the first few days of their captivity."

Life in the German hospitals could be tough. In May 1916, a naval admiral and parliamentarian told a House of Lords debate, "I know one case of a man who had his leg badly shattered, and whose diet while in hospital consisted daily of one ounce of food. His breakfast was coffee, without sugar or milk; his dinner, very poor soup containing decayed cabbage and sometimes carrots, with black bread; his supper, poor soup again with black bread; and he had tea on Wednesdays and Saturdays." The prisoner was given no medicine and there was no help from nurses.

Inmates had to look after each other. During his time at the hospital, he was allowed only two baths, with no soap. He was in that place for a year.

Perhaps Ted Brown had a different experience. Or perhaps his reticence to talk openly about the matter signalled a need to suppress difficult memories. For the first month, he kept news of the bullet wound from his family and friends, telling them in letters only of the lesser injuries he had suffered as the aeroplane crashed. But after a while he stopped this pretence. As his condition worsened, and infection bedded itself deeply into his foot, he began to admit to his parents how much pain he was suffering. After one operation in late March, he wrote, "They don't hurt you any more than is necessary, but at times it seems as if a lot were necessary."

Nonetheless, by August 1916, he was considered sufficiently well recovered to leave the hospital and was moved into Germany's sprawling prison camp system. His first placement was the Gütersloh camp in north-western Germany. Then he was transferred eighty miles east to Clausthal, a camp hidden in thick mist under blankets of snow high in the Harz mountains.

Life was no easier in Germany's camps than in its hospitals. One Canadian soldier described his arrival at Clausthal in March 1917. He had endured a twenty-seven-hour journey in an unheated train travelling in sub-zero conditions without food. "When we reached Clausthal," the soldier remembered, "the snow was about a foot deep and it was still snowing. We were in awfully bad condition by this time, half frozen, so stiff that we could scarcely move at first, and tired out, yet the first news we got was that we had to carry our suit cases through the snow to the camp about two miles away." On arrival, the prisoners were berated by the camp commandant. Then they were each stripped naked and given an intimate and humiliating search. It was only at three o'clock in the morning that they were sent to their freezing huts, unfed and with no water. The soldier recalled, "The camp was swarming with rats and mice."

Food rations in the camps were scandalously meagre. Those who were imprisoned had to rely on aid parcels sent from Britain to survive. In one parliamentary debate on the treatment of British prisoners, a speaker said, "The rations were such that, were it not for the parcels, the men would starve." The parcels were often delayed, stolen from, or lost. One pilot imprisoned at Clausthal later said, "With a diet lacking in vitamins, small cuts turned septic. Most of us went about with bandaged fingers."

Like many British officers jailed in the German camps, Ted Brown tried periodically to escape. In these attempts, he failed. Weakened by his wounded foot, and suffering increasingly from ill health, Brown was in little shape to go on the run in Clausthal's cold, mountainous terrain. Those that did manage to escape from German camps would likely be recaptured, returned, and punished. But there was another way to escape the misery of German imprisonment. Wounded or sick prisoners might be selected for internment in neutral Switzerland.

It was a scheme that had been operating for about nine months by the time Brown arrived at Clausthal in the autumn of 1916. Prisoners who had been wounded, but who might nonetheless be able to take on war roles if repatriated home, could be released from the German camps and held in Switzerland. As the fighting wore on, the German authorities concluded there was little point keeping sick and injured prisoners in their own care. Medical specialists were in short supply and they had enough work on their hands treating German casualties. The new arrangements saw Switzerland provide a way out. It would ease pressure in the German camps, and it would prevent men like Brown from returning home and contributing to the British war effort.

There were two stages to the process. In the first, travelling commissions of Swiss doctors visited the camps to select potential internees from among the wounded prisoners. Many fell at this first hurdle. Their wounds might be deemed too well recovered, or not serious enough, to merit inclusion in the internment programme. Those who passed this first commission would be sent to an exchange station. There, a second medical board would assess the candidates. This one included German doctors alongside Swiss medics. Some would fail at this second stage. Lists of rejected prisoners due to be returned to the prison camps from the exchange stations were sometimes known as "guillotine lists." To get so close, and yet be no farther forward, was crushing.

But there would be some who were passed by both commissions. These lucky few would soon be out of the exchange station on their way to the fresh air and comfortable accommodation provided by Swiss authorities.

The Swiss commissioners were quick to pass Ted Brown when they visited Clausthal in the middle of November 1916. The damage to his foot was apparent to anyone. The doctors also diagnosed a heart condition. It was enough to see Brown added to the list for the second panel. But he remained sceptical that he would make it to the end of the process.

After the Swiss commissioners departed, life in Clausthal returned to its normal, mundane rhythms. Christmas was around the corner and the prisoners' thoughts turned to ways they could celebrate and keep their spirits up despite their confinement. Days turned to weeks, with no news of Brown's departure. As time went on, it seemed like he had been right to be sceptical that he would leave the camp. But on the afternoon of December 14, a month after the Swiss commission's visit, he received word from one of the camp interpreters that he would be leaving. His destination was to be the exchange station at Constance, a German town on the border with Switzerland, 300 miles to the south. He would be leaving at five o'clock that afternoon.

Having hastily packed up his belongings, Brown reported to the camp's commanders and met the handful of other prisoners who would be travelling with him by train to Constance. Two were British. Percival Lowe was a captain in the West Yorkshire Regiment who had been captured in the first weeks of the war. The commission found him to be suffering from a nervous condition. Alongside Lowe was Harvey Frost, a Royal Flying Corps observer who had been shot down in a B.E.2c in April that year, suffering from a machine-gun bullet wound in his thigh. Frost had only been held at Clausthal since September. Lowe had been imprisoned there since August the previous year.

The journey to Constance was long and uncomfortable, with thirteen changes of train along the way. The party only arrived at their destination at midday on December 16, after almost two days travelling.

The medical commission visited the next morning. Men had been gathering from camps all over Germany, each with his own sad tale of injury and sickness. There were men with missing limbs and eyes; soldiers suffering from paralysis; those experiencing the disordering effects of trauma inflicted in the intense fighting of the trenches; and those, like Brown, who had multiple injuries and health problems. All passed in turn before the Swiss and German doctors before being returned to their accommodation. Nobody had any idea whether they had been successful or not.

The wait for results could be excruciating. A Canadian officer, Harvey Douglas, was sent to the Constance exchange camp a month earlier than Brown. He recalled his experience there in a letter home, which he sent soon after arriving in Switzerland. "I wrote you from Constance but you probably never got the letters," he said. "We were there for a month and the whole stay was nothing but a series of disappointments." Douglas described the feeling when fellow officers found they were to be refused entry into Switzerland. "Suddenly," he told his family, "as a bolt from the blue, a Guillotine list was read out containing the names of eight senior officers who were to go back to camp. It was some sort of reprisal, I am sure. We saw the Commission and then no one knew whether he had passed or not."

Douglas explained what happened when the Swiss medical commission arrived. "We were examined; notes were taken about our cases and we left the room without being told whether we had been accepted or refused. . . . One day an Unteroffizier came in with a list of six or seven names. This was read out and the officers mentioned were told to pack up and leave for camp at once. Almost every day a similar list came in and each of us trembled for fear his name might be included."

The findings of the commission held on December 17 were released after a wait of only four days. Brown, Lowe, and Frost waited anxiously as the results were read out. Two British officers who had come from the Heidelberg prison camp just before Brown and the rest of the Clausthal party arrived were told that their applications had been rejected. They were being sent back. As the German officer worked his way down the list, Brown, Lowe, and Frost got more and more nervous. If they were rejected, it would mean a return to stifling captivity. There was no end to the war in sight. Such a rejection might well have felt like a death sentence.

Eventually, the German officer reading out the list reached each of the three Clausthal prisoners in turn. They had all been passed.

There was still another wait. For four weeks, the prisoners remaining at Constance were repeatedly told that they "would be leaving in 5 days," as Percival Lowe wrote in his diary. Then the message given to the men by camp officials changed. Lowe recalled, "After that we were honestly told we were being kept for reprisals." Harvey Frost was later interviewed by a British government committee investigating the treatment of prisoners of war. In his testimony, he said, "We were informed that this delay was a measure of reprisal for the delay in the sending from England of some Germans who had passed the Swiss Commission." For Ted Brown, the frustration of being held within sight of Switzerland was hard to stomach.

The day at last arrived when Brown and his two Clausthal colleagues were to leave. It was January 18, 1917. After being driven in hospital lorries to the nearby railway station, the men and their baggage were locked in a room while everyone else arrived. Then they were loaded onto a waiting train. They were directed to different carriages depending on their destination. Brown, Lowe, and Frost had been allocated places in Switzerland's Château d'Oex region, 130 miles west.

Nobody could celebrate while the train remained stationary. Harvey Douglas, the Canadian officer who had gone through the process a month before Brown, later wrote, "There wasn't a cheer or any other sign of joy. We were still on German territory and before we crossed the border a wheel might come off the engine or there might be an earthquake."

But the wheels on Brown's train stayed on. After what seemed like a lifetime, the guard blew his whistle. Slowly, lurching, the train began to pull away from Constance station. Moments later, it came to the border between Germany and Switzerland. Lowe recorded the experience in his diary. "In a very short time I saw the German & Swiss sentry boxes on the side of the line," he wrote. "Then the Swiss doctor shook hands and said now you are in Switzerland. A sigh of relief here was not I think out of place." Douglas recalled his own departure: "We looked out of the windows and saw women and children standing in the snow outside their doors, waving handkerchiefs and shouting, 'Vive l'Angleterre.' A spontaneous cheer went up from every man. It was the most thrilling moment of our lives."

After twenty-four hours on the train as it passed slowly through Switzerland, and after a total of fourteen months of imprisonment, Ted Brown arrived in the village of Rossinière, near Château d'Oex. He had been released from German captivity and was now an internee. He wasn't home. But he was safe.

It turned out that he had got out of Clausthal just in time. Harold Medlicott, his pilot, wasn't so lucky.

A FEW MONTHS AFTER BROWN WAS RELEASED FROM CLAUSTHAL, Medlicott arrived there to take his place. Having been uninjured in the aeroplane crash, he spent the following eighteen months making a series of escape attempts from the camps in which he was imprisoned. Clausthal was his seventh camp.

Between Brown leaving and Medlicott arriving, a new commander of the camp group that included Clausthal had been appointed. Karl von Hänisch was a cruel and irrational man. He brought to the camps a programme of vicious intimidation of British captives, whom he hated. One prisoner reported that Hänisch operated a "reign of terror" and "appears to be mad." He said, "He will march, surrounded by his staff, and shriek, quivering with rage, at the British officers." It seemed like he was possessed. Another prisoner described him as "the most ghastly butcher dressed up as a German it is possible to imagine."

Hänisch's programme was enacted with brutal enthusiasm by the local commandants he brought in to run his camps from day to day. Clausthal's new head was Heinrich Niemeyer. Niemeyer once gathered his guards in earshot of British prisoners and shouted, "Here you have under you English prisoners. Take your revenge! Shoot them on every occasion! I am behind you! General von Hänisch is behind you! Your

Fatherland is behind you!" He meant what he said. He was soon known to order the indiscriminate bayoneting of British officers.

Once, following an escape attempt, Niemeyer had three young prisoners brought before him. He ordered sentries to strip the officers of their clothes. In a wild fury, he grabbed a necktie, wrapped it around one of the men's necks, and began to strangle him. Then he turned to another of the men and screamed, "You shall go down on your knees to me." When the officer hesitated for a moment, a sentry forced him, still naked, down to the ground.

Harold Medlicott wasn't prepared to put up with such humiliating treatment. In November 1917, after months enduring the violent behaviour of Hänisch and Niemeyer at Clausthal, he escaped. He was quickly recaptured and sent to nearby Holzminden, another camp under Hänisch's command. Holzminden was run by Karl Niemeyer, a brother of Heinrich. About seventy years old, he was described by one prisoner as a "bloated, pompous, crawling individual," and by another as "the biggest liar in Germany." One evening, a twenty-year-old British private, George Dellar, left his room to visit the latrines. He ran into Karl Niemeyer in the corridor. The commandant, suspicious, demanded to know where Dellar was going. When Dellar told him, Niemeyer flew into a rage. He shouted at the young soldier, "You can shit where you like: you can shit yourself: you can shit the bloody bed, but you will not go to the latrine tonight."

The conditions at Clausthal and Holzminden under Hänisch and the Niemeyers were considered the worst in the German prisoner-of-war camp system. A report prepared for a British government committee investigating Hänisch described the situation as "a harrowing catalogue of oppression, punishment and injustice."

Medlicott escaped from Holzminden in February 1918. He was recaptured and moved to Bad Colberg, a camp in central Germany. In May, he escaped from there as well. Once more, he was recaptured. This time, as he was being returned to the prison, he was shot dead by German troops. He was twenty-five years old.

A member of the Royal Flying Corps ground crew who prepared

Medlicott's aircraft the day he and Brown were shot down later wrote, "Had we but known what the Fates had in store!"

TED BROWN SPENT EIGHT MONTHS INTERNED IN SWITZERLAND. IT felt like a lifetime. Conditions were much better in the internment village than in the German prison camps and hospitals. Brown later admitted, "If one had not wanted to be back and in things again, one would have had a very jolly time." But his experience of imprisonment cast a lingering shadow. He told a *New York Times* journalist who interviewed him in 1919 that he would never fully recover from the twenty-three months he was held in captivity. In Switzerland, he was restless, nearly driven mad with boredom, and finding it hard to cope with the after-effects of trauma.

Eventually, though, his incarceration came to an end. On August 24, 1917, Brown wrote from Rossinière to a contact at the American legation in Bern. He told his friend that he had been passed by a medical commission for repatriation and expected to be released soon. Just over a fortnight later, in September 1917, Brown was back in Britain. He was a free man.

Brown used his two years in captivity to keep his mind active, even if his body was broken. He later observed, "It was as a prisoner of war that I first found time to begin a careful study of the possibilities of aerial navigation." He was still in pain from his injuries. But he, unlike Medlicott, was now free. His horizons had been cruelly closed off for two long years. Now they expanded once more. He had the opportunity to study, to perform experiments, and to converse at will with his old engineering friends.

REPORTS OF HAROLD MEDLICOTT'S SHOOTING BY GERMAN TROOPS near Bad Colberg began to appear in British newspapers on June 13, 1918. Nobody on the British side believed the German story that Medlicott had been shot trying to escape. A secret British government memorandum

The scene of Harold Medlicott's escape attempt from Holzminden camp in 1918.

observed, "The latest reports from Bad Colberg indicate that the officers there suspect foul play about this business." Rumours abounded among British airmen that Medlicott's body had been seen by a British soldier when it was brought back to the camp. The soldier reported that it was covered in bayonet wounds. It was commonly accepted that Medlicott had been murdered in cold blood by vengeful German soldiers.

Brown set off for the Royal Air Force base in Lincolnshire three days after those newspaper reports first appeared. Taking up a new role as a pilot instructor was a decision he had made the previous month. But learning of his partner's murder—and he must have seen the reports in *The Times*, the *Daily Telegraph*, and all the rest—could only have steeled his resolve. He and Medlicott had been snatched from the air by a hail of shellfire and bullets. They had been captured at the point of a rifle. Medlicott had refused to submit to his resulting captivity and had now paid the price with his life.

For Brown, the pain and humiliation of the crash and his subsequent imprisonment simmered for two and a half years. Six months working at the Ministry of Munitions office job in London had left the matter unresolved in his mind. Pilot instruction appeared to offer some sort of salvation, a sense of closure. But the war ended—and Medlicott's life had been taken—before Brown's mind could heal. Then the chance to fly the Atlantic with Jack Alcock came up. Alcock and Medlicott were the same age as each other. In the Vimy, with Alcock piloting, perhaps Brown could finish his journey in triumph, not tragedy. There were no German lines to navigate over the Atlantic. There was no gunfire and hostile shelling from below. It must have felt like a bet worth taking.

Brown had climbed into the Vimy in Lester's Field on June 14, 1919, filled with confidence. Yet here he was, as in 1915, about to crash. As he sat with Alcock in the cockpit of the aeroplane as it hurtled downward, he must have believed that his luck, like Medlicott's, had finally run out.

CHAPTER 20

THE ROTTENEST LUCK IN THE WORLD

THE NEEDLE ON THE ANEROID GAUGE FLICKERED BELOW 100 FEET. The heavy Vimy biplane, caught in a spiral dive halfway across the Atlantic, was about to hit the water. Maybe Jack Alcock and Ted Brown whispered prayers at that moment. Or perhaps they just shut their minds to what was about to happen. But as suddenly as the aircraft had entered the cloud at the top, it left it at the bottom. They had emerged from its disorienting embrace. They were still heading towards the ocean, but they could see again.

Immediately, Alcock's brain focused. He quickly glanced around, searching for external references that would allow him to recover the aircraft. He found the horizon. Because of the angle the Vimy was making at that moment, it looked as if the sea was standing up sideways. Ten years of flying had programmed Alcock to act almost automatically. As he processed what he saw, he adjusted the control column and rudder bar. The Vimy was responsive for such a large machine. Under Alcock's deft control, the aircraft came out of the dive and levelled off. Brown looked at the aneroid. It read fifty feet. He noted, "It appeared as if we could stretch downward and almost touch the great whitecaps that crested the surface. With the motors shut off we could actually *hear* the voice of the cheated ocean as its waves swelled, broke, and swelled again." Immediately, Alcock operated the throttle control to his side. The engines powered back to life.

As the aircraft recovered, Alcock and Brown saw from the compass needles that they were now heading west. Alcock brought the Vimy round in a wide arc and pointed it east. Then he pushed it into a climb away from the water that had almost claimed them for its own.

About forty minutes after they had entered the cloud, the airmen were back at 6,200 feet and searching for the sun. The dive had defrosted the air-speed indicator, which now worked normally. Brown had recovered his composure enough to offer Alcock a sandwich and make a sardonic comment on the notepad.

Alcock kept climbing. They were travelling above the lower layer of clouds through misty vapour. Their visibility was limited to a few yards beyond their wingtips. Above them, the sun had risen on a new day. But apart from a lightening of the clouds, they could see nothing new.

Then they hit bad weather. First, heavy rain came at them almost horizontally, pelting the windscreen and drenching the doped canvas of the aircraft. After a while, the rain turned into snow, which gathered on the wings and the struts and wires. Then the snow became sleet and was accompanied by a hailstorm, which stung and stabbed at the aircraft, and at any part of the airmen that was raised above the protective limits of the windscreen.

With no observations possible until they climbed above the clouds to the sun, Brown took a break from navigation and picked up the headphones of the wireless receiver. He wanted to listen for incoming messages. He had done this several times during the journey. Three hours into the flight, he heard a transmission that was signalling a ship's call sign. It was the slightest contact imaginable with the outside world: just three letters in staccato Morse code, directed at somebody else. It lifted Brown's heart to hear it. But his elation didn't last. He discerned a further transmission a few hours later, but it was indistinct. He tried a few more times after that. There was an occasional sound of static or interference, but nothing more. He noted, "The complete absence of such contact made it seem that nobody cared a darn about us."

Soon after 5 a.m., having climbed through thick cloud, heavy snow, and sleet, the Vimy reached the height of 8,800 feet. Every now and then, both Brown and Alcock glanced back at the petrol overflow glass on the central strut, looking for reassurance that fuel was flowing into the engines. Brown did so again now. This time, he found that the gauge had become covered by hard-packed snow. Alcock had to keep his hands and feet on the controls, so the uncomfortable job of clearing the snow off the gauge fell to Brown. He twisted round in his seat, then stood up on the floor of the cockpit facing backward. No longer shielded from the biting wind, he had to struggle to keep his footing. While Alcock kept the aircraft level, Brown grabbed one of the thick wooden struts and climbed up onto the seat cushion, before bracing himself with a knee on top of the fuselage. He took out a pocket knife, reached up, and carefully chipped the snow from the gauge. After clearing the glass with a gloved thumb, he saw that the fuel was passing safely through. Then he let himself down from the fuselage onto the seat, turned to face the front, and sat back down, grateful to be back within the protection of the cockpit. He would repeat this operation several more times as the snowstorm continued.

There was another safety task that needed to be carried out every so often. Brown would take out a battery flashlight, pop his head up again, and check the oil pressure and engine speed gauges mounted on the sides of the nacelles. Alcock knew more than most people what could happen if an engine in one of these big bombers failed. It could end in a nightmare, and it was only six months since he had woken up from his own.

IT WAS THE EVENING OF SEPTEMBER 30, 1917, JUST HOURS AFTER the dogfight over the Aegean that earned Jack Alcock his Distinguished Service Cross. The plan was to make another night attack on Constantinople in the Handley Page bomber. The targets this time would be railway infrastructure: they would be aiming for the main station at Constantinople and one at nearby Haidar Pasha. The last time the Turkish capital had been raided by the Handley Page, back in July,

Kenneth Savory had been in the pilot's seat. This time, Alcock would fly the plane.

Accompanying Alcock for the evening's bombing mission would be two of his most valued Mudros colleagues. Hugh Aird was a Canadian aviator who had passed through Alcock's training school at Eastchurch the previous year. The pair transferred to Mudros just three months apart. At twenty-five years old, Aird was the same age as Alcock and had matured into a skilled and respected pilot at the Aegean airbase. Then there was Stanley Wise, a motor mechanic from Bedfordshire, north of London. At thirty-one, Wise was a little older than Alcock and Aird, and he was an expert in the maintenance of Rolls-Royce engines. He had arrived at Mudros a couple of months earlier and accompanied Alcock on the successful night raid on Adrianople.

The Mudros ground crew loaded the Handley Page with thirty-two bombs for the mission. Twenty were high-explosive bombs, each weighing 112 pounds. A further twelve, lighter in weight, were incendiaries. Together, the payload weighed a ton. After supervising the loading, Alcock, Aird, and Wise climbed into the cockpit. They completed the final checks for the mission. Satisfied that everything was in order, Alcock shouted an instruction to the ground crew to start the engines. After they had warmed up, Alcock called for the chocks to be pulled away and then taxied the bomber down the Mudros airstrip. They were airborne at about 8.15 p.m. and headed north-eastward.

Alcock anticipated a journey of about three hours. Their 200-mile route would take them past the neighbouring island of Imbros before reaching the Gallipoli Peninsula. Then, clearing Gallipoli, it would be a long flight over the Sea of Marmara before arriving at Constantinople, which they hoped to reach by midnight. It was a route the airmen of Mudros now knew well.

It would be a challenging flight. The rarefied air of the region meant the aeroplane could fly no higher than about 6,500 feet. Stanley Wise later recalled flying through valleys lined by hills towering hundreds of feet above them: "The close proximity of 'Archie' under these conditions was very unpleasant," he observed.

An hour and a half into the flight, having reached the far end of the Gallipoli Peninsula, Alcock was about halfway to the target. Everything seemed to be going to plan, despite the anti-aircraft fire being directed towards them. But as they flew onward in the cool evening breeze, a sudden grinding metallic shriek was heard from one of the wing assemblies directly alongside the three crewmembers' heads. The aeroplane juddered and rocked as if it had been hit by an artillery shell.

Fixed to each of the aircraft's two engines was an installation known as the reduction gear. This epicyclic gearbox reduced the speed of the rotating propellers to their optimum levels while allowing the engine itself to rotate at a higher speed, maximizing its power output. It was a crucial component of any high-powered engine driving a large or relatively slow aeroplane, but it was one that had to endure great stresses. As Alcock had driven the heavy bomber along the Sea of Marmara, the reduction gear in the port engine exploded, shearing the installation from the body of the engine and sending the propeller attached to it tumbling away from the aircraft.

There was now no way the men could continue their bombing mission to Constantinople. But they still had one functioning engine; they might be able to make it home. After all, the same aircraft had been flown all the way back from the Turkish capital on one engine after the July raid. On that occasion, the craft had been much lighter by the time the engine failed. This time, the Handley Page still carried its ton of bombs and an almost full load of fuel.

Pulling the bomber carefully around in an arc, Alcock began to nurse it back over the Gallipoli Peninsula towards the open sea, aiming it towards the safety of their Mudros base.

It was a hard slog. The Handley Page had already demanded much of its two engines in getting as far as it did. Nonetheless, Alcock persevered. For almost an hour, the aeroplane and its three occupants flew on over the enemy territory of Gallipoli. Eventually, the western end of the peninsula could be glimpsed in the darkness. The wide expanse of the Aegean was in their sights, with Mudros dead ahead.

But the aircraft had been losing altitude all the way on the return journey. With thin air already limiting the machine's height, the single engine failed to keep it on a level path. Alcock realized he wouldn't be able to pilot it all the way back to Mudros. In fact, he wouldn't make it out of Turkish territory. After flying sixty miles on one engine, their altitude had dropped from 6,000 feet to just 1,500 feet. They were now over the water, and Alcock began to look for somewhere to ditch.

Wise and Aird began throwing the aircraft's maps, bomb sights, and machine guns over the side of the cockpit. Alcock started nosing down towards Suvla Bay, an inlet on the north-western tip of the peninsula. As they approached the water, the trio braced themselves for the landing, but it offered little protection when the moment came. Despite Alcock's careful handling on the approach, the aeroplane tipped forward, hard, as its wheels touched the surface. Then it plunged nose first into the cold sea.

The three men were thrown violently out of the cockpit. Later, Wise recalled his experience: "With a red mist in front of my eyes, and visions of many stars mingled with the sea foam, I found myself under water fighting frantically for breath," he said. After a long struggle, he broke free from the wreckage that was holding him under and reached the surface. As he did so, he spotted Alcock and Aird doing the same.

It seemed as if luck was looking out for the British airmen. Despite the damage it had sustained, the bomber remained afloat. The three men managed to climb back into its cockpit. Alcock fired off flares with his signal pistol in the hope of alerting the British destroyer that was stationed a few miles to the south. But the ship was too far down the coastline for its crew to see them. For two hours, Alcock fired intermittently into the night sky, praying that the ship's crew would spot them. Nobody came.

Instead, Turkish troops at a camp on the Suvla Bay coastline began shooting at the trio as they sat in the aircraft. At first, the shots were wildly off the mark. Soon, spotting where their bullets hit the water, the Turkish snipers adjusted their range. The gunfire rapidly closed in on Alcock and his crewmates. Then the aircraft began to sink.

The three airmen were under fire and had nothing left to cling onto. They climbed out of the cockpit, dropped quietly into the water, and began to swim for the shore. It took them an hour to reach the rocks that lined Suvla Bay. They silently hauled themselves out of the water under the protective cover of darkness, found a hiding place among the rocks, and waited. Sleep came to none of them.

As the sun began to rise on October 1, they began searching for something they could eat. But there was nothing to be found amid the barren rocks that hid them. The hours passed slowly. They were soaked through, shivering from the cold, hungry, and exhausted. Hope began to fade. At about ten o'clock that morning, they emerged from their hiding positions, fired their two remaining flares into the sky, and surrendered to three Turkish soldiers who quickly came to the scene.

THEIR CAPTURE WAS A HUMILIATING EXPERIENCE. THE FIRST thing the Turkish troops did was to strip them of their clothes. Their possessions—what little they managed to cram into their pockets as they abandoned the sinking aeroplane—were also confiscated. Then Alcock and his crewmates were pushed by their captors at bayonet-point towards the local camp in what Alcock later described as "somewhat embarrassing circumstances."

It was a five-mile march to the Turkish encampment. As they approached, fear began to grip them. Mudros airmen had been bombing the Suvla Bay camp for several months. On July 15, a formation of nine Sopwith aircraft from Mudros had flown over the bay and dropped a total of twenty bombs. The raid left the camp ablaze, with its storage sheds badly damaged. Alcock himself was likely one of the Sopwith pilots that day. On August 4, the day that Alcock was piloting the Handley Page to bomb Panderma, Mudros airmen bombed the Suvla Bay camp two further times, causing a fierce and deep-seated fire. Five days after that, in retaliation for a German attack on Imbros, British aircraft again bombed Suvla Bay.

This helped explain the degrading treatment meted out by the Turkish soldiers as they marched their captives to the camp. If one of the men fell behind, a soldier would strike him over the shoulders with his rifle.

To make matters worse, the path fell steeply away in places. The soldiers delighted in shoving their captives towards the edge of the precipice as they walked along.

Once the party arrived at the encampment, the Turkish commander came out to meet his new prisoners. To their relief, he greeted them with courtesy. Their clothes and possessions were returned to them. They were given a meal and allowed to rest. But the Suvla Bay camp was not intended for detention. Later that day, the three men were marched to a larger military base, where they were interrogated by Turkish and German officials. Then, still exhausted from their ordeal, the men were transferred 200 miles to the Turkish capital, Constantinople. It had been their original destination in the Handley Page bomber. Now, the city was being reached in changed circumstances. The tables had been turned on the British airmen.

Their destination was the Seraskerat city prison, a jail notorious for its brutality. It was supposed to be a temporary stop while arrangements could be made for transfer to an internment camp. Alcock and his colleagues were appalled by the conditions they confronted in the prison, where they were placed in almost solitary confinement. Alcock later recalled, "The food was scarce and bad—indeed, we looked like being starved—but what upset us more than anything else was the filthy condition of our temporary home. Vermin swarmed in every hole and cranny; I never wish to live under similar conditions again."

Francis Yeats-Brown, a Royal Flying Corps airman who also spent time imprisoned in Seraskerat, wrote about his experience of arriving there. He was escorted down into darkness. Then he was led along a corridor, hearing the screams of prisoners. A heavy iron gate was unlocked. He was pushed through. Then it was slammed shut and locked behind him. He recalled, "There were cries and shouts down there, and men scrambling for food, and other men who looked like wild animals, behind bars. . . . Well-fed rats were scurrying amongst the garbage, and badly-fed prisoners were pacing the room forlornly, or twiddling their thumbs, or scratching themselves, or gnawing crusts of bread." The underground room was filthy and stifling. He saw prisoners held in heavy chains, shackled at the wrists and ankles. Some had been kept like that for months or years. The men were emaciated and infested. His own bed consisted of two planks of wood on an iron frame. The wood was crawling with lice and bugs. He was given no bedding.

Yeats-Brown soon despaired of his new surroundings. "One was utterly lost in that dungeon," he recalled. "Even when the war ended, would one be found? I doubted it." He spent a month in solitary confinement. He described it as "a punishment as barbarous and as senseless as the thumbscrew or the rack: more so indeed, for it is better to kill the body than to maim the mind." A window in his cell looked onto a room sometimes used as a court. Once, he watched a police officer kick an elderly woman to the ground in front of two judges trying her case. Then the officer beat her repeatedly about the head using a wooden club as thick as his wrist. After she passed out, he began kicking her in the face. Sometimes, curtains were drawn across the window. Yeats-Brown could hear the screams of prisoners in the room being tortured. In the prison's garden was a gallows. He later said, "Captivity is a minor form of death, and I was dead, to all intents and purposes."

For Alcock and his two colleagues, it was a miserable, frustrating experience. Every day, the airmen expected to be moved out of the jail and sent to a detention camp. Each morning, their hopeful enquiries were rebuffed by the Turkish prison wardens, who would always tell them, "Tomorrow."

Two weeks into his incarceration at Seraskerat, Alcock wrote a letter to his uncle back in England. "I had the rottenest luck in the world," he said. "I would give anything to be at Mudros, being prisoner of war does not agree with me although I am treated very kindly & get plenty of food." It is possible that Alcock and his fellow captives were treated better than other prisoners in the jail. More likely is that their jailers told them what to write. Alcock would also have wanted to hide the true nature of the situation from his family. He was certainly doing his best to keep his crewmates' spirits up as they endured the first weeks of confinement. Stanley Wise, more than Alcock or Aird, had been greatly troubled by the experience, and he suffered a breakdown in the prison.

Years later, he wrote how Alcock's "smiling face and cheerful personality under these adverse circumstances" stuck in his fractured memory of the episode.

In 1902, when he was ten years old, Jack Alcock sat in his school classroom in St. Anne's-on-the-Sea, carefully copying out an essay into his exercise book. It was about courage, and its lessons must have shaped the character he carried through life. "Physical courage enables us to meet dangers or pain without flinching," he had copied. "Moral courage induces us to do right under all circumstances without fear of ridicule or punishment. Naturally timorous people may grow more courageous by using their reason. True courage is not mere foolhardiness or bravado; it is never associated with bragging or bullying. Physical courage is a noble quality when used on the side of right or in the service of humanity. Moral courage enables us to resist temptation and ennobles our nature." He would have needed deep stores of courage as he endured the incarceration of that Turkish jail.

After a dispiriting month at Seraskerat, the three airmen were finally transferred to a prisoner-of-war camp. They arrived on November 3, 1917. Their new home was Kedos, a town located about 140 miles south of Constantinople, halfway between the Black Sea and the Mediterranean. The main prison camp was positioned just to the north of the town on an acre of land. Some inmates were housed in rented buildings in the town itself as the camp grew. Alcock and his two colleagues joined about a hundred other detainees at the camp, including numerous Scottish and Indian prisoners. "Well old boy," Alcock wrote to his brother-in-law soon after arriving, "here I am doing time all through a burst propellor."

At Kedos, a parole agreement had been settled between the captives and their Turkish guards. It was a camp from which prisoners would not attempt to escape. In return for such assurances, detainees enjoyed more freedom than elsewhere. Alcock joined his fellow prisoners in amateur dramatic productions. They were permitted to go on walking and climbing excursions in the surrounding countryside. Some would sit by the nearby river and sketch, or swim, or collect flowers. Sometimes, imprisoned officers would go out with Turkish farmers to hunt wild boar in the hills or sit and pass the time with the town's shopkeepers. It was a far cry from the fetid town jail in Constantinople. Alcock later wrote, "There we had a good time, with freedom and abundant food. It was indeed a happy place, that camp at Kedos."

But Alcock could never forget that he was in captivity. The sleeping arrangements at the camp were crude and uncomfortable. Sanitation was basic. Food was plentiful but expensive, and the cost only increased as the months of his incarceration drifted by. He passed the time as best he could, but the boredom ground him down and he missed his friends. "I wish I could get amongst the fun again this life is driving me dotty," he wrote to a fellow flyer back on Mudros, James Sinclair. He added, "Tell all the boys to write and send snaps." A fellow Kedos prisoner recalled that Alcock was "a dedicated airman, a true professional, who kept himself to himself and had little interest in anything but flying and longed to return to it." Alcock was affable and popular, but never really gregarious. Once, during a walk in the nearby hills, he left his friends to walk by himself for a while. Suddenly, he was set upon by a pack of aggressive German shepherd–like dogs, which dragged him to the ground in a frenzy of bites and snarls. Luckily, his companions were close enough to see the attack. They rushed over with sticks to beat and kick away the dogs, freeing Alcock from their teeth. That time, he escaped with bite wounds to his arms and legs. His friends told him that he might have been killed. He carried a three-foot truncheon on walks from then on.

In September 1918, ten months after Alcock and his colleagues arrived at Kedos, life was disrupted once more. That month, an outbreak of influenza spread rapidly across the camp. Then, on September 27, as the prisoners were recovering from their illnesses, a house fire in one of the town's small wood-and-mud dwellings quickly spread to other buildings, fanned by a strong breeze. A dry summer had left water supplies low, making attempts to attack the fire futile. The layout of the town, with its narrow and tightly packed streets, meant the flames quickly took hold. Before long, it was an inferno.

The entire camp population raced to the scene of the fire. For hours, they fought the flames. They spent the rest of the night attempting to rescue property and create firebreaks. Nobody was killed. But when the sun rose the next morning, with weary British prisoners still labouring to salvage property, the scene was one of devastation. Most of the town, including many of the properties housing British detainees, had burned to the ground. One officer said, "Kedos had ceased to exist. Out of some 2,300 houses, more than two thousand had in a few hours been reduced to ashes, and in the daylight the ruined town showed a scene of extraordinary desolation." Another prisoner lost all his possessions to the fire: his personal diaries, precious letters from home, irreplaceable photographs. He described Kedos as "a town of the dead." Jack Alcock later said, "None of us will ever forget the sight."

After the fire had been put out and the smoke began to clear, many of the Kedos prisoners found themselves homeless. Some managed to cram into the buildings of the main camp. But many were forced to sleep outside, with no protection from the elements. Alcock was one of the unlucky ones. "We lived in the open for a month," he recalled. With winter coming, the prisoners prayed for good weather, but October saw the arrival of a cold and wet snap. To add to the men's hardship, food supplies became hard to find. Much food had been burned in the fire, and what little was left was taken by residents fleeing to neighbouring villages. It was a miserable time for the Kedos prisoners.

Then, it was all over. In November, world events overtook matters. The war ended.

Plans to move the Kedos prisoners to a new camp after the fire had long since dissolved and they had been left to fend for themselves. Their final trek started with a journey from Kedos to Smyrna, on the country's Aegean coast. There, they would await ships to be repatriated. On November 18, Alcock and other Kedos prisoners finally left Turkey for the long and tortuous journey home via Egypt. Alcock arrived in England on December 16, 1918.

It had been fourteen and a half months since he was forced to ditch the Handley Page bomber on the waters of Suvla Bay. Even while incarcerated, his optimism couldn't be stifled. It quickly grew into defiance. If there was a challenge, he would meet it. According to Stanley Wise, it was in the Seraskerat prison that Alcock first voiced his desire to compete in the transatlantic contest. By the time he got to Kedos, the idea had become a certainty. "He told me in Kedos," recalled a fellow detainee, "that he would be the first man to fly the Atlantic."

CHAPTER 21

WE DID NOT LET YOU DOWN

IN THE SKIES OVER THE ATLANTIC OCEAN, JACK ALCOCK PUSHED the Vimy ever higher in search of the sun. Sleet and snow continued to cover the aircraft. By 6.20 a.m. they were at 9,400 feet. The tops of the wings were sheets of ice and the radiator shutters in front of the engines had frozen. Packed snow had also frozen in the aileron hinges, jamming the huge flaps solid. For an hour, Alcock had no lateral control of the aeroplane. Luckily, the rudders and elevators still worked, and the Vimy was an inherently stable craft. Alcock cruised onward and upward with caution.

At seven o'clock, the pair reached 11,000 feet. The clouds above them thinned and briefly broke. Brown was able to take a quick sighting of the sun. After he computed their position, he found that they were close to Ireland, but north of their intended track. He suggested to Alcock that they descend so that the machine could warm up. They might also spot a steamship that could help guide them on their way. Alcock nosed the aircraft gently downward.

As he did so, the starboard engine began popping back through a carburetor. It sounded ominous. Brown figured that a broken or sticky valve might be causing the backfires. Alcock throttled the engine back and the noise subsided.

Slowly, they glided down. After forty minutes of gentle descent, they had reached 1,000 feet. The ice in the aileron hinges had now melted, so they were working again. But the aircraft was engulfed once more in clouds. The airmen couldn't see anything beneath them. The aneroid showed their altitude, but it wasn't always exactly right. It was affected by changing atmospheric conditions as well as height. There was a risk when descending through clouds that the ocean might be closer than they expected. Alcock was making the shallowest descent, just in case. If they hit water like this, there was a chance they would be alright, but it was nerve-racking.

This time, they came out under the cloud layer at 500 feet, and saw the dull, grey ocean spread out beneath their wings. Alcock gunned the throttles again and both engines jumped back to life. The popping had stopped; a short rest seemed to have cured the starboard motor. Brown picked up his drift indicator and took a sighting off the waves. They were still north of their route, but not so far north that they would miss Ireland altogether. At 8 a.m., he asked Alcock to steer south-southeast.

To keep clear of the lowest clouds, Alcock needed to fly at just two or three hundred feet above the ocean. There were no ships, there was no colour in the sea, no brightness. Just grey waves. Brown reached back to the provisions cupboard and brought out another couple of sandwiches and some chocolate, though he and Alcock were less hungry than ever. They had nearly reached land—they hoped.

At 8.15 a.m., Brown was putting the wrappings back into the cupboard behind him. He never forgot what happened next. "Alcock grabbed my shoulder, twisted me round, beamed excitedly, and pointed ahead and below," he remembered.

It was still raining. The view was murky. But Brown could see what Alcock had spotted. Below them were two green islands which lie just off the coast of Ireland—Eeshal and Turbot—and ahead of them were white breakers. Brown scribbled a short note in his navigator's logbook: "In sight of land."

At 8.25 a.m., they crossed the coast.

After a few minutes looking for landmarks, they spotted the tall wireless masts of the huge Marconi station at Clifden, County Galway. They

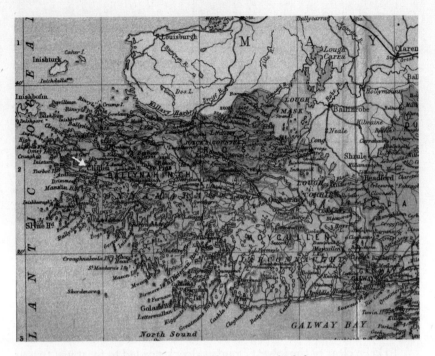

Map of Galway. Clifden is just over halfway up on the left.

were exactly halfway down Ireland's western coast. Brown's navigation had been spot on.

Brown fired two red flares into the air to attract attention, but nobody saw them. Alcock piloted the Vimy towards Clifden town, two miles from the wireless station, and circled for a while. Then he started to search for a landing spot. He found an open space near the Marconi buildings which looked like it would do the job. He throttled the engines off, turned into the wind, and slowly glided the Vimy down to the ground. At the very last moment, he flattened out, and the aeroplane touched down. It was 8.40 a.m. on Sunday, June 15, 1919.

As the wheels contacted the ground, they rolled forward at first. Then, without warning, they started to dig in. Rapidly, the nose of the Vimy tilted forward while the tail lifted into the air. The ground was a bog; it was too soft to support the heavy aircraft.

Just in time, Brown braced himself against the cockpit wall. Alcock used his feet on the rudder bar to ready himself for impact. By force of habit, he reached out and switched off the engine magnetos, killing the ignition sparks. Then the aircraft came to an abrupt halt.

The Vimy's nose, its lower wing planes, and its twin propellers were buried in the soft bog. The nose, made of a tubular-steel framework covered in doped fabric, buckled. But it held, protecting the cockpit. Brown bumped his face in the impact. Aside from that, neither airman was hurt. However, the fuel lines linking the petrol tanks to the carburetors had snapped in the crash. There was enough petrol on board to fly for another ten hours, and fuel began spewing out of the tanks into the cockpit. Quickly, the airmen grabbed the mailbag, logbook, notepad, and instruments. Then they climbed out and moved clear. Once they were safely away, Brown said to Alcock, "What do you think of that for fancy navigating?"

Alcock replied, "Very good." Then he shook Brown by the hand.

After the pair collected themselves, Brown fired off two white distress flares into the sky, signalling for help. Within minutes, a party of officers, soldiers, and operators based at the wireless station came labouring across the boggy ground.

An officer shouted out, "Anybody hurt?"

"No," the airmen answered.

"Where you from?"

"America."

ALCOCK AND BROWN SPENT THE MORNING AT THE WIRELESS STATION. They sent off telegrams: to family; to Vickers at Brooklands; to Rolls-Royce; to the Royal Aero Club, running the contest; and to the *Daily Mail*. They also cabled the mechanics back at St. John's with a simple message: "Your hard work and splendid efforts have been amply rewarded. We did not let you down." Then they bathed and had breakfast.

That afternoon, a twenty-seven-year-old *Daily Mail* reporter, James Hodson, arrived at the wireless station. He found Alcock and Brown standing over a large bag, quietly stowing their flying suits and instruments. As he reached them, he offered his congratulations. Alcock

Ted Brown and Jack Alcock at Clifden wireless station soon after landing.

looked up and grinned. Brown, his eyes a little bloodshot, bent further over the bag.

Hodson was in a hurry to get a story away in time for the day's copy deadline. He escorted the airmen to the officers' mess, sat them down, and began to ask questions. He later observed, "Naturally enough, they amplified one another, they corrected one another, they interrupted one another, until, in despair, we all agreed it would be best if each told his own story."

Alcock went first, and the experiences of the sixteen-hour flight came tumbling out. When he came to recall the spiral dive, he said, "We came down quickly from four thousand feet until we saw the water very near. That period lasted only a few seconds, but it seemed ages. It came to an end with us within fifty feet of the water and with the machine practically on its back."

By five o'clock, Hodson had finished his interview with Alcock and raced off down the dirt path to the post office at Clifden to wire the copy to his editor. A couple of hours later, Alcock and Brown joined him there. The three men got into a car provided by the Marconi station and set off on the two-hour drive to Galway. Along the way, Hodson asked Brown to tell his side of the story. For an hour, shouting over the noise of the car as it bumped and ground its way along the road, Brown told his tale of star sightings, drift measurements, and course computations. But it was as if he was in a dream. "I answered questions mechanically," he later wrote, "and wanted to avoid the effort of talk." He told Hodson, "We are too near it to realize what it is we have done."

Brown had lost his grip on time and felt separated from reality. As his words came out, his mind was recreating moments from the flight. He could hear the rhythmic pounding of the engines. He could see clouds all around. He replayed, over and again, the spiral dive to the ocean. His mind showed him visions of what he had seen in the final moments of the dive: "the sudden sight of the white-capped ocean at the end of it." Alcock sat silently throughout.

After the interview was over, Hodson climbed into the front seat to urge the driver to go faster, leaving Alcock and Brown in their daze in the back. The car twisted along the road to Galway. The dark Connemara mountains were lost in deep shadows and thickening mist. The car passed crowds of onlookers who had gathered by the roadside in the rain. News had spread fast. Along the way, they encountered the Royal Aero Club official who was heading for the wireless station to verify the aircraft's arrival. At 9 p.m. the party reached Galway. James Hodson left for the post office to send off a longer report. For Alcock and Brown, a party had been organized. But it was clear to everyone that the airmen were dead on their feet. They had been awake for forty hours. The sixteen-hour flight had involved the most intense mental and physical concentration. Then they'd had to recalibrate their existence from complete isolation to the presence of great, cheering crowds. The party was over by ten o'clock. The two men went off to a nearby hotel to sleep.

They awoke at 7 a.m., ate breakfast, then entered another whirl of gatherings and adulatory crowds. At a big civic reception that lunchtime, they were asked to say a few words. Halfway through his second faltering sentence, Alcock broke down from the emotion of it all and couldn't continue talking. The crowd cheered to cover his embarrassment, then Brown stepped in with his own short speech.

From Galway they were taken by train through soft mist and rain to Dublin. Alcock placed Lucky Jim, his big-headed black-cat mascot retrieved from the Vimy's tail cupboard, on the seat next to him. It had done its job. At every station along the way, well-wishers had gathered. At Mullingar, a military band, accompanied by troops from the local regiment, welcomed the train into the station. A local schoolboy presented Alcock with a little wooden aeroplane he had made. It was a re-run of what had happened at Lester's Field just before they set off. Alcock said to the boy, "Thanks very much, old chap. A jolly good bit of work."

The scene at Dublin was one of jubilation. At the Automobile Club, grinning onlookers waited while Alcock and Brown sat patiently for photographs. Alcock held Lucky Jim gently in his lap. Brown carried the model aeroplane from Mullingar. Between them was the mailbag, with Twinkletoe placed carefully on top. The airmen spent the night in Dublin, and took the ferry to Holyhead, in Wales, the next morning.

Then it was another journey. This time, to London. Along the way, at Crewe, Alcock and Brown got out to greet crowds. Soldiers and porters hoisted the men onto their shoulders. Back in the train, Alcock said, "I didn't like it much, but I enjoyed Brown's face."

Kathleen joined the train at Rugby, reuniting with Ted for the first time in a month and a half. When news of the flight's success had reached the Kennedy home, a delighted Kathleen hugged her father, then kissed her mother, then danced around the room with her sister. A newspaper reporter had just arrived at the house. She told him, "Teddy's safe. He's done it. He said he wanted to so much, and I guess this is the proudest moment of his life." That morning, in the dreamlike moment between sleep and wakefulness, Kathleen had watched the floral pattern of her bedroom wallpaper rearrange itself into the shape of a horseshoe. "I regarded it as a lucky omen," she said, "and sure enough it has been." There was as much crying as dancing in the Kennedy home that day.

Two aeroplanes from Hendon, flying low, escorted Alcock and Brown's

Ted Brown and Jack Alcock being driven through London crowds to the Royal Aero Club.

train into the capital. There were more crowds and cheers at Euston station. Harry and Muriel Hawker were there to greet the party, but they couldn't get through the throng to reach either Alcock or Brown. But Muriel spotted Kathleen. She grabbed her and escorted her to the Hawkers' big Sunbeam motor car. Harry joined them and they drove Kathleen through packed London streets to the Royal Aero Club for the big reception. The open-topped car carrying Alcock and Brown led the procession. People shouted after them, "Good old Alcock!" and "Good old Brown!"

At the club, there was pandemonium. That evening, the precious airmail that Alcock and Brown had flown all the way from Newfoundland was delivered to its disbelieving recipients. Brown went to the Kennedy home in Ealing with Kathleen, where he faced yet another reception and more crowds, though he seemed to enjoy the attention. Alcock went to see Joe Beckett beat Frank Goddard in the heavyweight title fight at Olympia. Alcock got almost as big a cheer when he walked in as the two boxers did.

The next day, Alcock and Brown headed to Brooklands. The Vickers workforce had been more nervous than most as they waited for news of the airmen's fate that weekend. Annie Boultwood, who ran the structural fabric shop, remembered Archie Knight finally bringing them the news. He told her, "Mate, we've done it."

"Thank God for that," she replied.

To celebrate the achievement, Vickers management had laid out a reception at the works. Alcock and Brown each stood up in turn and thanked the men and women who had built the Vimy. They told the workers how their labour had stood up to the toughest test: a Vickers aeroplane, built by their hands, had made the first nonstop flight across the Atlantic Ocean. Annie Boultwood, Archie Knight, and all the other Vickers workers raised the roof with their cheers.

A few weeks later, Jack Alcock caught up with Bob Dicker. They sat at the edge of the Brooklands racetrack and talked about the flight. Alcock confided in his old friend how he had felt when the Vimy buried its nose at Clifden. At first, as the wheels began to sink, he thought the undercarriage axle they'd hastily repaired just before taking off had failed. Then he realized they were on bogland. It was his biggest regret. His plan had always been to land in Ireland, check the aircraft, then take off right away for Brooklands, so that the workforce there could celebrate the Vimy's homecoming. But the bog put paid to the plan. The idea that he had let the workers down upset him. "Bob," he told Dicker, "I could have cried."

THE *DAILY MAIL* THREW THE BIGGEST PARTY. ON FRIDAY, JUNE 20, Alcock and Brown were guests of honour at a luncheon at the Savoy Hotel in London. Lord Northcliffe was recovering from surgery on his throat and couldn't attend. The newspaper's editor, Thomas Marlowe, presided instead. There were more than three hundred guests that day— politicians and newsmen, industrialists and commentators. There were armed-forces leaders, aviators, and diplomats, from both sides of the Atlantic. H. G. Wells was there.

To Alcock and Brown, most of the guests would have been strangers. But there were a few in the room they knew well and would have been glad to see. Harry Hawker attended, as did Charles Morgan, still getting over his injuries. Alcock's old Brooklands friend Archie Knight was one of the men representing Vickers. The Sunbeam designer Louis Coatalen sat among the crowded tables. The aviation journalist Charles Grey was there as well. Only a few months earlier, Alcock had been corresponding with Grey from Turkish captivity. And there was family as well, or family soon-to-be. David Kennedy, Kathleen's father, came from Ealing, and two men travelled from Manchester to attend: Arthur George Brown and John Alcock, the airmen's fathers.

The Alcock and Brown family homes in Manchester had been mobbed on the Sunday afternoon when news of the men's success started to break. For a while, all Mary Alcock could tell reporters was, "I am too overjoyed to say anything." Later, she said, "We were certain he would do it, because all along he has been so confident. He was home about six weeks ago, and almost the last words he said were, 'Have no fear whatever, and don't worry. I am certain to do it.'" Arthur Brown had been more phlegmatic. Speaking outside a door flanked by British and American flags, he told journalists he had not even realized the flight had started until that morning, and by afternoon he knew of its success. "So you see," he explained, "there was very little time in which to get uneasy."

At the Savoy, it fell to Winston Churchill, secretary of state for war and air, to present the *Daily Mail* cheque for £10,000. Towards the end of his speech, Churchill paused. Then, unexpectedly, he announced that both airmen were to receive knighthoods from the king. The audience stood as one and cheered, but embarrassment flashed over Alcock and Brown's faces. They had no idea this was coming. Beforehand, during the lunch, Churchill had revealed there would be "something for him," Alcock said later. He thought it would be the Air Force Cross, an award for courageous service while flying. The idea of a knighthood, he confessed, left him "swamped."

The speeches the two men gave in response were characteristically modest and, in Alcock's case, visibly emotional. They spent as much time praising the Vickers and Rolls-Royce workforces as they spent talking about their own achievement.

Winston Churchill presenting the *Daily Mail* cheque to Jack Alcock and Ted Brown.

The American ambassador, in his response, lauded the sporting spirit that had characterized the contest. He also remarked approvingly on the Anglo-American nature of the winning team. Woodrow Wilson cabled Alcock personally to offer congratulations to him and Brown. The fact that Brown was an American citizen delighted US reporters. One described "a most happy augury for Anglo-American friendship that the first non-stop Transatlantic flight should be an Anglo-American effort." Northcliffe, though absent from the luncheon that day, had already described what Alcock and Brown's achievement promised. It was a typical newsman's angle: a world where London's morning newspapers would be selling in New York that evening, and the New York evening editions would be in London the next morning.

Brown, writing later, praised Northcliffe for his support of aviation in its infant years through the mounting of prize competitions. "In each case," he said, "the competitions seemed impossible of fulfilment at the time when they were inaugurated; and in each case the unimaginative began with scoffing doubts and ended with wondering praise. Naturally, the prizes were offered before they could be won, for they were intended to stimulate effort and development. This object was achieved."

The day after the *Daily Mail* lunch, just a week after they had taken off from Lester's Field to start the journey of their lives, Alcock and Brown went to Windsor Castle to be knighted by the king. In Newfoundland, some reporters were amused to hear the news. One wrote, "The thought of Alcock, in particular, being addressed as 'Sir John' brings a smile to those who best know the unaffected, democratic simplicity of this 27-year-old veteran pilot." One of Alcock's fellow aviators, on hearing of the flight's success, said, "I'm glad old Jack has got it. He's such a damned good chap."

The following month, the two men were honoured at a civic reception in their hometown of Manchester. More than a thousand people packed the state rooms of the town hall to receive the guests. As the pair walked in, the band played "See, the Conquering Hero Comes." A few days later, the aviators attended an event at the Vickers factory in Crayford. Prince Albert, later to become King George VI, opened the company's theatre, newly rebuilt after a wartime fire. After the prince had given a speech, Vickers directors brought Alcock and Brown onto the stage. They looked out over a sea of faces in the theatre's auditorium and offered a few words of thanks to the company for the opportunity they had been given. Then each of the two knights of the air was presented with a gold wristwatch as a memento.

ON SEPTEMBER 20, KATHLEEN AND TED BROWN BOARDED THE Cunard liner *Mauretania* at Southampton and headed across the Atlantic on their honeymoon. Twinkletoe, the black-cat toy that had last crossed the ocean in the Vickers Vimy, was stowed safely in their trunk.

The pair had married on July 29, a few days after the Vickers event in Crayford. The ceremony took place in the Savoy Chapel, just off Strand in central London. Sightseers eager for a glimpse of the now-famous airman crowded the streets outside the church for hours beforehand. But when Ted arrived, nobody recognized him, not even the police officers guarding the gate. Then Jack Alcock turned up, and he certainly caught the crowd's attention. He was mobbed. Cheers bounced around the buildings of Savoy Street as he walked through the gate to the churchyard. Reporters said he got to the door of the chapel "blushing like a schoolboy."

Inside, the opening hymn began with the words "Lead us, heavenly Father, lead us o'er the world's tempestuous sea."

Arthur Rostron was still captaining *Mauretania* when the Browns made the crossing to New York. He asked Ted about the transatlantic flight three months earlier. Brown said, "I wouldn't do it again for anything on earth."

AFTER LANDING IN IRELAND, JACK ALCOCK HAD TOLD ONE reporter, "We were tired of being alone in the fogs and the wilderness of sea." It might have been a metaphor for his war, a recent experience which still weighed on him. Later that autumn, he invited a dozen of his wartime flying friends to a drinks and dinner party at his home near

Kathleen and Ted Brown on board *Mauretania* on their way to New York.

Brooklands. He described it as a "binge." "I want you to quite understand that it is going to be a comic affair," he wrote in his invitation to James Sinclair, the fellow Mudros aviator who had kept him supplied with cheery letters in Kedos during his dark days of incarceration. "I have quite a lot of sour wine to get rid of, as well as some hot air," Alcock said. He promised Sinclair that the house would be filled with "all the noisy scamps I can get hold of."

ON THE STREETS OF NEWFOUNDLAND, THE NEWS THAT ALCOCK and Brown had landed safely was greeted with rapture. After so many months of gruelling slog; after so many disappointments and frustrations; after so long waiting for favourable winds, and for snow to melt, and for lease negotiations to be transacted; and after the near tragedies of Hawker, Grieve, and Morgan, at last the job had been done. The public was overjoyed. Fred Raynham's open-handedness in offering his airfield, his hangar, and his spare fuel to Alcock was remembered and praised. The two aviators, who had been friends since they were teenagers, had looked out for one another. To Newfoundlanders, that counted for a lot.

But if the citizens of St. John's were ecstatic, the aviators remaining there were less demonstrative. The best that local reporters could get out of one of them was the comment, "The boys did jolly well. They must have had bally good weather all the way. 127 miles an hour. That is really some going." Nonetheless, it was clear they were thrilled. Raynham was quick to praise the airmen, and quick to say his own efforts would be redoubled to follow in Alcock's footsteps. At Harbour Grace, the Handley Page team offered their congratulations as well. Herbert Brackley, the twenty-four-year-old pilot, was the most effusive. It was only four years since Jack Alcock had taught him to fly.

Once Alcock and Brown had taken off in the Vimy on June 14, the Handley Page men abandoned their attempt to get away the same day with their improvised radiator repair. There was no point taking the risk any more. Instead, they waited for the replacement radiators to arrive from England. On June 18, Brackley took the Handley Page up for a further test flight. A fuel load of 1,100 gallons had been filtered into the tanks, and six passengers, including Robin Reid, joined Brackley on board. The take-off run was short: only 245 yards. During the flight, Brackley circled at length over the airfield, keeping a close eye on the water temperature gauges. To his relief, they stayed cool: the new radiators were doing their job. After an hour cruising the skies over Harbour Grace, he brought the heavy craft successfully down to land. The flight was a complete success. But by this point, the weather had worsened and the favourable westerly winds over the Atlantic had stopped blowing.

The next day, rumours were circulating that Handley Page would switch their flight destination from Europe to New York. The firm's US manager, William Workman, was supporting the scheme. He was eager to capitalize on the publicity that surrounded the giant aeroplane. Mark Kerr pushed back. He said such a move would be "giving the American people a chance to look at a machine that had not done its job." But he was overruled. Two weeks later, he received a telegram from Handley Page in London. It instructed him to fly to America, not Ireland. According to Herbert Brackley, the message said that "no useful purpose would be served in attempting the Atlantic after Alcock's successful crossing." After all their preparation, they would never even try the Big Hop. Kerr later observed, "It was a bitter disappointment to have to give up the flight." Now that Handley Page had withdrawn from the contest, only Martinsyde was left.

CONRAD BIDDLECOMBE ARRIVED ON THE EVENING OF JUNE 14, just a few hours after Alcock and Brown made their getaway. Since then, the Martinsyde airmen and their mechanics had laboured to complete work on the Chimera, first at the Blue Puttee dance hall, then back in the canvas hangar at Pleasantville. Once the aeroplane had been assembled, they made a test flight. Apart from small snags, it went well. After that, it had been a wait for the weather, just like in May. On July 17, two days after Raynham's twenty-sixth birthday, the reports looked good enough to go.

The sky over Pleasantville that afternoon was a clear, vivid blue. The sun beat down brightly on the scarlet-and-yellow-painted biplane as it stood by the shore of Quidi Vidi Lake. Crowds had come to the aerodrome; not so many as last time, but enough to make a festive atmosphere for the departure. The people of St. John's had taken Raynham into their hearts. His youthful determination appealed to them. He had not allowed himself to be set back after the crash that put Morgan out of the running. The young pilot had picked himself up, found a spirit of leadership he did not know was within him, and brought the aircraft back to life. Now was his chance to prove himself.

Raynham and Biddlecombe chatted happily with friends and spectators as the Falcon engine was started. After it had warmed up, Raynham climbed into the cockpit, followed by Biddlecombe. The two men settled into their seats and made final adjustments to instruments and settings. Then Raynham called for the chocks to be pulled from the wheels. He opened the throttle to full, and the Martinsyde aeroplane began to bounce along the same rutted turf that had brought the Raymor to grief two months earlier. For 300 yards the Chimera headed into the westerly wind. The roar from its engine echoed across the hills lining the valley. It gathered speed as it rolled and lumbered across the ground. Then, with Raynham's deft control of the stick, it lifted off the ground.

The crowd, their faces having been masks of anxiety during the takeoff run, relaxed into joyful celebration. The aeroplane rose to thirty feet, its wings casting raking shadows onto the ground as it flew onward and upward. As Raynham nursed the heavy machine into the sky, Biddlecombe settled in and prepared for the journey ahead. The delighted onlookers, watching from vantage points around the airfield, thrilled to the sight. It was the third successful departure of a transatlantic flight from their city. Their cheers filled the air.

Then a gust of crosswind blew momentarily across the field and lifted the tip of the aircraft's right wing. The machine pivoted around its fuselage. Because the aeroplane was still so low, the left wingtip grazed the rough ground. As it dragged, it pulled the entire aircraft around to the left, turning the machine almost ninety degrees, until the Chimera was facing due south and heading for the lake. Raynham struggled to regain control and level the craft. It was too heavy. He didn't have enough lift. Then it plummeted thirty feet and smashed into the ground. The undercarriage collapsed under the impact and the nose and engine ploughed a jagged path into the turf until the whole thing lurched to a halt. The propeller had been shattered. The lower wing planes were ripped and crumpled. Supporting struts had splintered and snapped and the bracing wires were now a knotted tangle of steel. The fuel tank, filled for the transatlantic crossing, had split in the crash. Petrol began to gush out.

Onlookers rushed forward to help. Raynham and Biddlecombe were sitting motionless in their seats. The stench from the fuel coursing out of the tank and soaking the wreckage hit the noses of those closest to the crash. Before anybody could get to the aircraft, Biddlecombe reached up, pushed aside the wires and struts above his head, and climbed slowly from the aircraft. Then Raynham swung a long leg over the side, pulled himself out of his seat, and climbed down as well.

Martinsyde officials raced across the airfield with fire extinguishers. Biddlecombe, seeing Raynham's ashen face, left him to his thoughts for a moment and tried to prevent people from getting too close.

It had happened again. Raynham just stood there, unable to speak. He was staring at the crumpled Chimera in disbelief, his mouth trembling with emotion, oblivious to the shouts from onlookers. Struggling to keep tears at bay, he stalked around the aircraft, taking in the damage it had sustained. He turned to the Rolls-Royce engineer who had just arrived at the scene and said, "Do you want your engine? I'm through with my part and if you are—well, just touch a match to it." The shock and the frustration rolled across him in waves.

Reporters approached, asking what had happened. He told them that if the crosswind had hit the airfield two seconds later—just two seconds later—he would have had enough height and speed to make it away. But there was nothing he could have done.

Then the reporters asked if he would carry on. Raynham waved his hand towards the splintered ruin of wood and fabric and wire, and the

The crashed Chimera.

broken engine, and the ground soaked with petrol, and replied, "I have nothing left to 'carry on' with."

FRED RAYNHAM WAS BACK AT THE MARTINSYDE FACTORY IN England by the end of July. Soon after his return, six of the seven transatlantic airmen held a reunion dinner. It was a chance to remember the time they spent together at the Cochrane Hotel in Newfoundland. They were a small band, those dauntless aviators who had taken off from St. John's attempting to make the Big Hop. Seated around a big table at the Café Royal on London's Regent Street were Harry Hawker and Mac Grieve; Fred Raynham and Conrad Biddlecombe; and Jack Alcock and Ted Brown, the pair who made it all the way. Only Charles Morgan, who was still recovering from his injuries, was absent. There were others at the dinner who had been at St. John's as well, including Montague Fenn, the Sopwith manager; Percy Maxwell-Muller, from Vickers; and the Royal Air Force meteorologist, Lawrance Clements.

The men dressed up for the occasion. Some were in dinner jackets, others in smart suits. Brown and Clements wore their Royal Air Force uniforms. But it was no buttoned-up, starchy affair. In the elegant surroundings of the restaurant that evening, they all relaxed and let off a little steam. The menu card included a set of joke rules that recalled the fun they'd had at the Cochrane together. There was to be no throwing of bread; no water fights. They were smiling and happy as they dined that evening. As it happened, this would be the last time the airmen would ever sit around a table together. But that night, they all enjoyed the time they spent in each other's company. Doubtless, they talked excitedly of the future.

After the meal, they stepped out of the restaurant into the cool autumn air. They said their goodbyes, and then each of the men went his own separate way.

Clockwise from left: Jack Alcock, Eric Platford (Rolls-Royce specialist), Harry Hawker, Donald Woods Mason (lubrication specialist), Montague Fenn, Conrad Biddlecombe, Ted Brown, Mac Grieve, Lawrance Clements, Percy Maxwell-Muller, and Fred Raynham.

EPILOGUE

THOUGH THE SEVEN AIRMEN OF 1919 ARE ALL NOW LONG GONE,
there is one character from that episode still here and in good health.
After Jack Alcock and Ted Brown left Clifden, the Vimy aeroplane was
dug out of the bog, sent back to the Vickers factory at Brooklands, and
restored. On December 15, 1919, six months to the day after Alcock
landed it on the Galway coast, it was presented to the British nation and
put on display at the London Science Museum.

For twenty years the Vimy sat in pride of place among the aeronautical exhibits on Exhibition Road. When the Second World War broke out,
it was taken away for safekeeping. After the museum reopened, it was
displayed in a building on nearby Queen's Gate. Then, in the early 1960s,
it was given a new location: a vast aeronautics hall built on the top of the
museum to resemble an aircraft hangar, with arch girders supporting an
aluminium roof soaring thirty feet above the floor. It is still there today.
Along one side of the hall, showcases are filled with absorbing models
and relics recounting the story of flight from its earliest days. Along the
other side stands industrial racking crammed with engines. Occupying
the main space of the hangar are the aircraft. There are two dozen of
them. Some are suspended from the ceiling as if in flight. Others stand
on plinths to afford a closer look.

Every year, three million people visit the Science Museum to explore

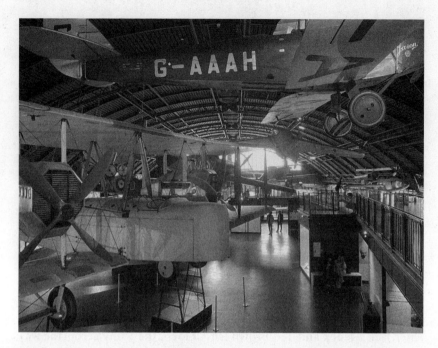

Science Museum aeronautics gallery, 2024, with the Vimy on the left and the aeroplane used by Amy Johnson in 1930 to fly to Australia suspended overhead.

the relics of the modern age. They come—many by aeroplane—from all over the world, and pass, excited, through the front doors and across the marble floor of the lobby. They wander past rows of nineteenth-century steam engines. Then they enter the ground-floor gallery which most recently housed space rockets and lunar landers, but which, in December 1919, was where the restored transatlantic Vimy first went on show.

Today's visitors can then take the elevators to the flight exhibit in the hangar on the top floor. There, they might join a guided tour given by docents or curators. Their guides will explain some of the highlights on show. They might point to the rudimentary little motors that drove some of the earliest aeroplanes, then to the big engines that once powered jet airliners, including one from a supersonic Concorde. They take visitors up onto the elevated walkway that runs the length of the hall, and point out the triplane made by Alliott Verdon-Roe in 1909, and the biplane that

Samuel Cody flew in 1912, and the German fighting monoplane from the early days of the Great War.

As the visitors wander with their guides through the exhibits, they might catch the faint scent of the historic aircraft around them: the smell of engine oil, and varnish, and fabric dope, smells that would be instantly familiar to the Brooklands pioneers. At the far end of the hall, guides sometimes linger beside three aircraft that took Harry Hawker's name: a 1930s Hawker Hurricane, built at Brooklands, and two 1960s Hawker jets. A slice through a Boeing 747 fuselage covers the end wall.

Then the guides escort their charges back along the walkway to the centre of the room, where the transatlantic Vimy stands. It is so large that its tail almost touches the side of the gallery, yet so fragile it seems like a miracle it ever flew. But it did. The guides show off the fuel tanks fitted along the length of the aeroplane that once got filled with gasoline in a muddy field in Newfoundland. They point to the petrol overflow glass, high on the strut above the fabric-covered fuselage, that became packed with hard snow a mile and a half above the Atlantic. They point out the two powerful engines planted between the wings, and they gesture at the cramped open cockpit. Children crane their necks to take it all in. Most of the adults do as well. Then the guides explain how Jack Alcock and Ted Brown sat in that very cockpit for sixteen hours in June 1919, flying the aeroplane nonstop across the Atlantic Ocean and into the future, and everybody listens in wonder.

I know this because, for many years, I was one of those guides. My first day on the Science Museum staff was September 11, 1995, when, as a twenty-one-year-old trainee, I was toured around the building by my new colleagues. I was captivated by everything that I experienced, but one hall stuck in my memory. It was the aeronautics gallery. I remember the day vividly. The theatrical vista of so many famous craft was a thrill. When the display was formally opened in 1963, the education minister, Sir Edward Boyle, said, "The ancients dreamed of attaining a god-like power, that they, like the Olympians, might span the skies and rival Mercury." Here, in front of me, was the material realization of that ancient dream.

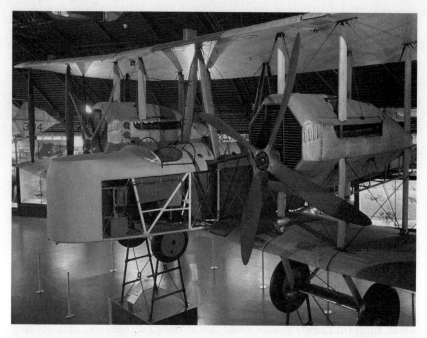

The transatlantic Vimy at the Science Museum, 2024.

Fifteen years after I joined as a trainee, I stood in the same gallery in my new role of transport curator. Now, these world-changing aircraft were my care and responsibility. One of them always called out to me: the Vimy. Yet, as I carried out my work, I came to find that few people seemed to know the story of 1919.

It wasn't as if it had been forgotten by those daring aviators who followed in the Vimy's wake. When Charles Lindbergh landed in Paris after his solo crossing from New York in 1927, he reportedly told onlookers, "Why all this fuss? Alcock and Brown showed us the way." Amelia Earhart, who flew the Atlantic alone in 1932, described Alcock and Brown's flight as "an amazing feat, and the least appreciated." Amy Johnson, after her solo journey from England to Australia in 1930, said she considered Alcock and Brown's flight "the greatest in the history of aviation." In 1934, Jean Batten repeated Johnson's feat, only faster. On a celebratory tour of her New Zealand homeland, Batten was hosted at a reception by Hilda Hewlett, then aged seventy, whose aviation school had taught Maurice Ducrocq to fly twenty-four years earlier. Without Hewlett, and without Ducrocq, Jack Alcock might never have got his Brooklands break.

In the public's eyes, though, the brilliance of Alcock and Brown's story had evidently faded. As my years at the museum passed, and I took ever more visitors around the aviation exhibits, the story became lodged in my heart. I wanted to bring it out from the shadows. I never met Jack, nor Ted, nor all the others who took part in that most audacious challenge in 1919; writing this book is my way of paying tribute.

One newspaper described performances like Alcock and Brown's flight as "tall lighthouses on the path of progress." But lights can dim or gutter out if they are not kept kindled. It would be twenty years after that first crossing, in 1939, that Pan American would inaugurate the first scheduled transatlantic air service, using a fleet of forty-ton, seventy-four-seat Boeing flying boats. Revolution can be slow for those living in it. But change does come, and it affects us all in the end. Charles Morgan, the last of the airmen of 1919 to die, learned to fly in a frail biplane made from sticks of spruce covered in linen. He lived long enough to see Concorde fly supersonic.

Today, a nonstop flight across the Atlantic is routine. Perhaps it even seems like a chore. But once, it was a journey into the unknown.

It was not inevitable that we would take to the skies; still less that we would fly across great oceans. It happened because people made it happen, and those people were adventurers of the truest kind. Today's routine transatlantic flight is only possible because of those adventurers, those romantics, those dreamers. We can fly today because of all the heretics who went before us: the men and women, young and old, from all walks of life, who felt an urge to step off their fated path; to achieve something that had not been done before; to reach the limits of possibility and pass far beyond them. We fly today because of those seemingly ordinary people who performed extraordinary acts, people who decided to change the course of their own lives, and in doing so changed the course of humanity.

The life stories of the seven transatlantic airmen after their Newfoundland adventure ended make for melancholy reading. There was a

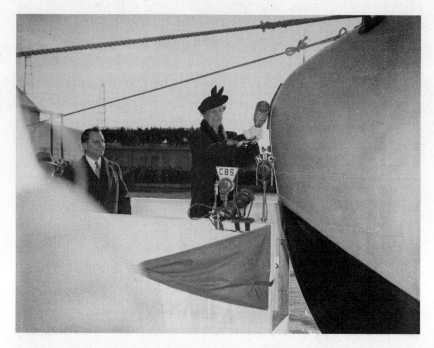

First Lady Eleanor Roosevelt christening Pan American's *Yankee Clipper* on March 3, 1939, the aircraft which flew the first scheduled transatlantic service.

lot of pain and heartache as the rest of their lives unfolded. But in the autumn of 1919, as six of them gathered around a Café Royal dinner table with their friends to celebrate what they had achieved, that was all yet to come; an unknowable future. At that restaurant table, on that cool London evening, they were modern heroes. Their eyes shone; their young faces grinned; their bodies sat alert and alive. They'd been in the game from the very jump. They'd risked everything for the Big Hop, and it paid off.

A nonstop flight across the Atlantic might be routine to us. But it is only possible because of those who went first.

ACKNOWLEDGMENTS

IT IS THIRTY YEARS SINCE I FIRST WALKED BENEATH THE CANVAS wings of the big Vickers Vimy at the Science Museum, looked up at its open cockpit, and wondered what must have possessed two young men to fly it across the Atlantic. Many people have guided me along the path I stepped onto that day.

May I first thank the staff of the libraries, archives, and museums who gave me advice or whose collections I used to help piece this story together. I have listed these institutions in a later section, and I am obliged to all the people who answered my calls. Special mention goes to the very helpful teams at three archive collections on Newfoundland: Memorial University of Newfoundland (MUN), the City of St. John's Archives, and the Archives Department at The Rooms. A particular thank you to Colleen Quigley at MUN for her guidance and enthusiastic support at a crucial stage of research.

For their invaluable specialist advice, I would like to thank Derek Bird, Tim Boon, Robert Bud, Roy Clare, Kristin Hibbs, Larry Hibbs, Molly Hibbs, John Liffen, Natasha McEnroe, and James Nye. In Crayford, where Vickers built aeroplanes, Janet Hearn-Gillham, David Gillham, and Barnaby Smith were enormously supportive. I'm also grateful to members of the High Wycombe Society and the Penn & Tylers Green Residents Society. For generously giving me access to archival material in their possession I am indebted to Annie MacDougall, granddaughter of James Sinclair; Guy Griffiths, great-nephew of William Allcock; and Gail Hewlett, granddaughter-in-law of Hilda Hewlett. My correspondence with Gail during the writing of this book was always uplifting and invigorating and I am most grateful for her warm encouragement.

Special thanks go to Tony and Rita Alcock. Tony is Jack Alcock's nephew and a highly distinguished aviator himself. Tony and Rita were most generous with their advice, their hospitality, and their support, and I am honoured to have spent time in their company.

Andrew Nahum, aviation historian and former senior keeper of the Science Museum, has been a friend, mentor, and inspiration throughout my museum career, and gave liberally of his knowledge on this project. I thank him most sincerely.

I have greatly benefited from the Research Associateship I hold at the Science Museum Group and the one I formerly held at Royal Holloway, University of London, for both of which I am grateful.

At W. W. Norton I thank John Glusman, Helen Thomaides, Claire Shang, Wick Hallos, and their many colleagues, who turned an ambition into reality and enabled this chapter of my life to be an utterly enjoyable one. At Chatto & Windus I am grateful to Clara Farmer, Rosanna Hildyard, and all of their colleagues for the same reasons. My agent, Jack Ramm at Aevitas, has been an unfailing guide not just throughout this project but since we first worked together. I greatly appreciate his counsel, which is always wise.

Finally, and with love, I thank my family.

David Rooney, London, 2025

SELECTED SOURCES

THE STORY OF THE BIG HOP RESIDES, FRACTURED INTO COUNTLESS fragments, in books, newspapers, reports, archive collections, photograph albums, film reels, and museum displays far too numerous to describe in full. I list here the repositories I used and the sources I relied on most heavily.

BOOKS AND ARTICLES

Accounts written by the transatlantic airmen themselves have been vital, if scanty. These include John Alcock, "My Transatlantic Flight," *The Badminton Magazine* 52, no. 290 (September 1919): 376–88; Arthur Whitten Brown, *Flying the Atlantic in Sixteen Hours* (New York: Frederick A. Stokes, 1920); and Harry Hawker and Kenneth Mackenzie Grieve, *Our Atlantic Attempt* (London: Methuen, 1919).

Then there are works written by those close to the action. Muriel Hawker wrote a biography of her husband in 1922, after he had died in a flying accident the previous year. It proved invaluable. See *H. G. Hawker, Airman: His Life and Work* (London: Hutchinson, 1922). Mark Kerr, of the Handley Page crew, wrote an autobiography that described his abortive activities in the transatlantic adventure. See *Land, Sea, and Air: Reminiscences of Mark Kerr* (London: Longmans, Green, 1927). Herbert Brackley's wife, Frida Brackley, wrote a memoir of her husband after his death in 1948. See *Brackles: Memoirs of a Pioneer of Civil Aviation* (London: Simpkin Marshall, 1952). A vivid account of the American transatlantic crossing in the NC aircraft was published as *The Flight across the Atlantic* (New York: Curtiss Aeroplane and Motor Corporation, 1919).

Over the years there have been several book-length accounts of the transatlantic contest and wider attempts to cross the great ocean by air. Among them, in chronological order, are: Charles Dixon, *The Conquest of the Atlantic by Air* (London: Sampson Low, Marston, 1931); Graham Wallace, *The Flight of Alcock & Brown* (London: Putnam, 1955); Basil Clarke, *Atlantic Adventure: A Complete History of Transatlantic Flight* (London: Allan Wingate, 1958); F. H. Ellis and E. M. Ellis, *Atlantic Air Conquest* (London: William Kimber, 1963); David Beaty, *The Water Jump: The Story of Transatlantic Flight* (London: Secker & Warburg, 1976); Percy Rowe, *The Great Atlantic Air Race* (London: Angus & Robertson, 1977); Gavin Will, *The Great Atlantic Air Race* (Dublin: O'Brien Press, 2011); and Brendan Lynch, *Yesterday We Were in America: Alcock and Brown, First to Fly the Atlantic Non-Stop* (Stroud: History Press, 2019).

Of the many works exploring the early years of British aviation, I kept returning to these: Claude Grahame-White and Harry Harper, *The Aeroplane: Past, Present, and Future* (Philadelphia: J. B. Lippincott, 1911); Charles C. Turner, *The Romance of Aeronautics* (Philadelphia: J. B. Lippincott, 1912); R. Dallas Brett, *The History of British Aviation 1908–1914*, vols. I and II (London: Aviation Book Club, c.1935); Brian Robinson, *Aviation in Manchester: A Short History* (Manchester: Royal Aeronautical Society Manchester Branch, 1977); Chris Aspin, *Dizzy Heights: The Story of Lancashire's First Flying Men* (Helmshore: Helmshore Local History Society, 1988); Michael H. Goodall, *Flying Start: Flying Schools and Clubs at Brooklands 1910–1939* (Weybridge: Brooklands Museum Trust, 1995); and Michael H. Goodall and Albert E. Tagg, *British Aircraft Before the Great War* (Atglen, PA: Schiffer Military History, 2001). The situation in Australia is explored in Horace Miller, *Early Birds* (London: Angus & Robertson, 1968). Helena Bryden, ed., *Wings: An Anthology of Flight* (London: Faber and Faber, 1942), is a wonderful international compendium.

Memoirs and biographical accounts of some of aviation's pioneering figures were particularly enlightening. See, for instance, F. Warren Merriam, *First through the Clouds* (London: B. T. Batsford, 1954); Graham Wallace, *Flying Witness: Harry Harper and the Golden Age of Aviation* (London: Putnam, 1958); Constance Babington Smith, *Testing Time: A Study of Man and Machine in the Test Flying Era* (London: Cassell and Company, 1961); L. K. Blackmore, *Hawker: One of Aviation's Greatest Names* (Shrewsbury: Airlife Publishing, 1990); Alan Bramson, *Pure Luck: The Authorized Biography of Sir Thomas Sopwith, 1888–1989* (Yeovil: Patrick Stephens, 1990); Gail Hewlett, *Old Bird: The Irrepressible Mrs Hewlett* (Leicester: Matador, 2010); and Stella Pixton, *Howard Pixton: Test Pilot and Pioneer Aviator* (Barnsley: Pen & Sword Aviation, 2021).

There are too many biographies of Alfred Harmsworth, Lord Northcliffe, to list here. I explored a wide selection. A recent and excellent addition to the field is Andrew Roberts, *The Chief: The Life of Lord Northcliffe, Britain's Greatest Press Baron* (London: Simon and Schuster, 2022).

To understand military aviation in the First World War and its aftermath, I drew from these works, among others: Harold Rosher, *In the Royal Naval Air Service: Being the War Letters of the Late Harold Rosher to His Family* (London: Chatto & Windus, 1916); Hilda Hewlett, *Our Flying Men* (Kettering: T. Beaty Hart, 1917); William Bishop, *Winged Warfare* (New York: George H. Doran, 1918); "Wing Adjutant" [Wilfred Blake], *The Royal Flying Corps in the War* (London: Cassell, 1918); Evan John David, *Aircraft: Its Development in War and Peace and Its Commercial Future* (New York: Charles Scribner's Sons, 1919); Charles C. Turner, *The Struggle in the Air, 1914–1918* (London: Edward Arnold, 1919); Elliott White Springs, ed., *War Birds: Diary of an Unknown Aviator* [John McGavock Grider] (London: John Hamilton, 1927); and Walter Raleigh (vol. 1) and Henry Jones (vols. 2 to 6), *The War in the Air: Being the Story of the Part Played in the Great War by the Royal Air Force* (Oxford: Oxford University Press, 1922–37).

The experiences of airmen and other combatants captured during the war were vividly expressed by accounts including these: J. Harvey Douglas, *Captured: Sixteen Months as a Prisoner of War* (New York:

George H. Doran, 1918); Francis Yeats-Brown, *Caught by the Turks* (London: Edward Arnold, 1919); Charles Leonard Woolley, ed., *From Kastamuni to Kedos: Being a Record of Experiences of Prisoners of War in Turkey, 1916–1918* (Oxford: Basil Blackwell, 1921); Tony Spackman, ed., *Captured at Kut: Prisoner of the Turks—The Great War Diaries of Colonel W. C. Spackman* (Barnsley: Pen & Sword Military, 2008); Neil Hanson, *Escape from Germany: The Greatest PoW Break-Out of the First World War* (London: Doubleday, 2011); and Oliver Wilkinson, *British Prisoners of War in First World War Germany* (Cambridge: Cambridge University Press, 2017).

A useful account of the British Westinghouse company and its successor, Metropolitan-Vickers, can be found in John Dummelow, *1899–1949* (Manchester: Metropolitan-Vickers, 1949). The history of Martinsyde is extensively described in J. M. Bruce, "A History of Martinsyde Aircraft," in the *Aeronautical Journal* (September 1968), 755–70; and Ray Sanger, *The Martinsyde File* (Tunbridge Wells: Air-Britain [Historians], 1999). Histories of other aircraft manufacturers can be found in the peerless Putnam Aeronautical Books series.

Newfoundland's colonial-era history is explored in, among many other titles, P. T. McGrath, *Newfoundland in 1911* (London: Whitehead, Morris, 1911); Lord Birkenhead, *The Story of Newfoundland* (London: Horace Marshall & Son, 1920); and Don C. Seitz, *The Great Island: Some Observations In and About the Crown Colony of Newfoundland* (New York and London: Century Co., 1926).

PHYSICAL REPOSITORIES

I made use of material held in: the Aeronautical Society of South Africa; BBC Written Archives Centre; Bodleian Libraries; British Library; Brooklands Museum; Cambridge University Library; City of St. John's Archives; the Common Room; Harper Adams University; the Institution of Engineering and Technology; Institute of Materials, Minerals and Mining; Institution of Mechanical Engineers; IWM London; London Library; The London Archives; Manchester Central Library; Memorial University of Newfoundland; National Aerospace Library; The

National Archives; National Collections Centre, Swindon; Parliamentary Archives; the Archives Department at The Rooms, Newfoundland; Royal Aero Club Trust; Royal Air Force Museum London; RNLI archive; Royal Meteorological Society; Science and Industry Museum, Manchester; Science Museum, London; Shell Historical Collection; Mitchell Library, State Library of New South Wales; Surrey History Centre; and University of Manchester Archives. The extensive and meticulously assembled Brian R. Robinson Aviation Archive at Manchester Central Library deserves special mention for those interested in early aviation in Britain. It is quite remarkable.

DIGITAL REPOSITORIES

I pieced together much of the story using digital resources provided by: Ancestry; the Aviation and Aerospace Archives Initiative; Aviation Ancestry; British Newspaper Archive; British Pathé; Center for Research Libraries; Espacenet; Europeana 1914–1918; Findmypast; Gale Primary Sources; *The Gazette*; Grace's Guide to British Industrial History; Hansard; HathiTrust Digital Library; International Committee of the Red Cross Archives; the Internet Archive; Kingston Aviation; Library of Congress; McCord Stewart Museum; Memorial University Digital Archives Initiative; National Library of New Zealand (Papers Past); *New York Times*; Oxford Dictionary of National Biography; the Probate Registry; ProQuest Historical Newspapers; Trove; and the World Newspaper Archive.

NOTES

PART ONE

7 **"They are heroes, modern ones"**: Hilda Hewlett, *Our Flying Men* (Kettering, UK: T. Beaty Hart, 1917), 37.

CHAPTER 1: FREEDOM AND EXHILARATION

9 **"crawling under machinery"**: *Lone Hand*, April 1914, 352.

9 **"Don't stand around doing nothing"**: L. K. Blackmore, *Hawker: One of Aviation's Greatest Names* (Shrewsbury, UK: Airlife Publishing, 1993), 32.

10 **"He thinks in gears"**: *The Herald* (Melbourne), May 20, 1919, 9.

11 **"As soon as I was up"**: *Queensland Times*, March 25, 1910, 2.

12 **"In his black Celtic eyes"**: *Daily Mail* (London), August 28, 1913, 5.

12 **He favoured actions over words**: *Penshurst Free Press*, March 24, 1911, 2.

12 **He was a keen and fast boxer**: *Sydney Mail*, July 20, 1921, 10.

12 **"he wanted more air"**: *Daily Mail* (London), August 28, 1913, 6.

12 **He was heading for England**: *Penshurst Free Press*, March 24, 1911, 2.

13 **In letters home, Hawker fumed**: Horace Miller, *Early Birds* (London: Angus & Robertson, 1968), 6.

13 **"They didn't think I was clever"**: Alan Bramson, *Pure Luck: The Authorized Biography of Sir Thomas Sopwith, 1888–1989* (Yeovil, UK: Patrick Stephens, 1990), 17.

14 **"All right, I will sweep"**: *Daily Mail* (London), July 13, 1921, 7.

14 **A friend recalled**: Miller, *Early Birds*, 11.

17 **"the immense flat plain"**: W. Boddy, *The Story of Brooklands: The World's First Motor Course*, vol. 1 (London: Grenville Publishing, 1948), 7.

18 **"I was rooted to the spot"**: Gail Hewlett, *Old Bird: The Irrepressible Mrs Hewlett* (Leicester: Matador, 2010), 109.

18 **"a passing and silly escapade"**: Hewlett, *Old Bird*, 127.

18   "The time will come": *The Tatler*, July 3, 1912, 34.

21   "His carefree looks": Geoffrey Dorman, "F. P. Raynham, O.B.E.," *Royal Aero Club Gazette* 8, no. 6 (June 1954): 198.

21   A government investigation in 1895: *Minutes of Evidence Taken before Her Majesty's Commissioners Appointed to Inquire into the Subject of Agricultural Depression* (London: Her Majesty's Stationery Office, 1896), 4: 116–17.

22   "It means," wrote the reformers: B. Seebohm Rowntree and May Kendall, *How the Labourer Lives: A Study of the Rural Labour Problem* (London, Edinburgh, and New York: Thomas Nelson & Sons, 1917), 312–13.

25   "amongst men, young and old": Hewlett, *Old Bird*, 159–60.

26   They dragged a hut door: *The Aero*, November 1912, 319.

28   It held this inscription: *Flight*, January 28, 1911, 1.

CHAPTER 2. ENGLAND IS NO LONGER AN ISLAND

29   "He seemed to be in a world": Paul Ferris, *The House of Northcliffe: The Harmsworths of Fleet Street* (London: Weidenfeld & Nicolson, 1971), 21.

30   Young Alfred would reportedly reply: Ferris, *House of Northcliffe*, 21.

30   "He had golden hair": Reginald Pound and Geoffrey Harmsworth, *Northcliffe* (London: Cassell & Company, 1959), 29.

32   "The Board Schools": Max Pemberton, *Lord Northcliffe: A Memoir* (London: Hodder & Stoughton, 1922), 29–30.

32   "Their standard of importance": Hamilton Fyfe, *Northcliffe: An Intimate Biography* (London: George Allen & Unwin, 1930), 84.

33   "Every day, there is an event": Pemberton, *Lord Northcliffe*, 63.

33   "Talking-points every day!": Fyfe, *Northcliffe*, 79–80.

33   "The mass of people have no tastes": Fyfe, *Northcliffe*, 65.

33   "remarkable new inventions": *Daily Mail* (London), May 4, 1896, 4.

33   "In his bed, with a secretary sitting": Fyfe, *Northcliffe*, 73–74.

34   "I never think of him": Tom Clarke, *Northcliffe in History: An Intimate Study of Press Power* (London: Hutchinson, 1950), 18.

34   After seeing reports of Santos-Dumont's: Clarke, *Northcliffe in History*, 96.

34   Charlotte Street, a thoroughfare: *John Bull*, January 25, 1930, 18.

35   "already knew that Germany": Pemberton, *Lord Northcliffe*, 63.

35   "the 'German Menace' provided him": Clarke, *Northcliffe in History*, 82.

36   "This news," Harmsworth said: Clarke, *Northcliffe in History*, 97.

36   "as he had made up his mind": Clarke, *Northcliffe in History*, 97.

36   "stimulate British airmanship": Clarke, *Northcliffe in History*, 97.

37   One experimenter who witnessed them: Graham Wallace, *Flying Witness: Harry Harper and the Golden Age of Aviation* (London: Putnam, 1958), 82.

37   "I have known to-day": Charles C. Turner, *The Romance of Aeronautics* (Philadelphia: J. B. Lippincott, 1912), 252–53.

39   "a very long way off": Wallace, *Flying Witness*, 96.

39   Northcliffe had previously lamented: Pemberton, *Lord Northcliffe*, 110.

39   **His aircraft was a delicate little monoplane:** *Daily Telegraph* (London), July 27, 1909, 11.

39   **"In two or three seconds I am safe":** *Daily Mail* (London), July 26, 1909, 7.

39   **"Do you realize it is the first time":** Louise Owen, *The Real Lord Northcliffe: Some Personal Recollections of a Private Secretary 1902–1922* (London: Cassell, 1922), 24.

39   **"The British people have hitherto":** *Daily Mail* (London), July 26, 1909, 6.

40   **One was David Lloyd George:** Wallace, *Flying Witness*, 127.

CHAPTER 3: ONE OF THE SOUNDEST PILOTS IN THE COUNTRY

43   **"He could make a superb parkin":** *Lancashire Life*, June 1969, 44.

44   **"I can well remember my sister":** John Alcock and Arthur Whitten Brown, *Our Transatlantic Flight* (London: William Kimber, 1969), 25.

44   **"Jack would make a very capable":** *Lancashire Life*, June 1969, 44.

44   **"They are not content":** Hansard: HC Deb, July 6, 1908, vol. 191, c1330.

45   **"for ever tinkering about":** *Belfast Telegraph*, June 16, 1919, 3.

47   **"There was never a book":** *Belfast Telegraph*, June 16, 1919, 3.

48   **"At Weybridge in 1911":** *Lancashire Life*, June 1969, 44.

48   **"sweating all day amid the oil":** *Belfast Telegraph*, June 16, 1919, 3.

49   **"Like a baby with a new rattle":** *Daily Telegraph* (London), August 4, 1911, 11.

49   **"The moment he started out":** *The Aeroplane*, August 10, 1911, 232.

50   **"Every day's work ends":** *The Aero*, May 31, 1910, 425.

50   **"Gradually one or another":** F. Warren Merriam, *First through the Clouds* (London: B. T. Batsford, 1954), 52 and 55.

50   **"no cocaine, morphia or laudanum":** Arnold Strode-Jackson, "The Fascination of Flying," in the *Aerial Year Book and Who's Who in the Air 1920* (London: Cross-Atlantic Newspaper Service, 1920), 36.

51   **"Jack Alcock, Brooklands, Weybridge":** Science and Industry Museum Archive: YA1997.36.10.

51   **"bombastic young Aussie":** *West Australian* (Perth), November 23, 1940, 7.

51   **"As soon as Hawker started":** Constance Babington Smith, *Testing Time: A Study of Man and Machine in the Test Flying Era* (London: Cassell, 1961), 66.

52   **"He was usually the first up":** R. Dallas Brett, *The History of British Aviation 1908–1914* (London: Aviation Book Club, 1935), 2: 99.

CHAPTER 4: THE GAME SEEMED GOOD

55   **"Although of American parentage":** Arthur Whitten Brown, *Flying the Atlantic in Sixteen Hours* (New York: Frederick A. Stokes, 1920), 8.

56   **Arthur's engineering obsession:** Clive Holland, "On the Ground, In the Air, and Over the Ocean: Some of My Many Thrills (Sir Arthur Whitten Brown)," *Motor Owner*, October 1, 1922, 46.

59   **"I had always longed":** Brown, *Flying the Atlantic in Sixteen Hours*, 8.

59   **"saw service in the trenches":** Brown, *Flying the Atlantic in Sixteen Hours*, 8.

60 **Recalling his early experiences:** "Wing Adjutant" [Wilfred Blake], *The Royal Flying Corps in the War* (London: Cassell, 1918), 2.

60 **"Mrs Hewlett's machines":** *The Star* (Christchurch, NZ), March 6, 1915, 3.

60 **"Nothing but raw material":** *Flight*, June 11, 1915, 419.

62 **"The observer is a trained":** "Wing Adjutant" [Wilfred Blake], *Royal Flying Corps*, 6.

62 **"It is no child's play":** William Bishop, *Winged Warfare* (New York: George H. Doran, 1918), 22.

64 **"immunity from death may seem":** "Wing Adjutant" [Wilfred Blake], *Royal Flying Corps*, 38.

64 **"God, it was a horrible sight":** Elliott White Springs, ed., *War Birds: Diary of an Unknown Aviator* [John McGavock Grider] (London: John Hamilton, 1927), 75.

64 **"a most skilful":** *Brecon and Radnor Express*, June 14, 1917, 7.

64 **"War is cruel":** Springs, *War Birds*, 36.

67 **"an insatiable appetite":** *Flight*, June 18, 1954, 808.

67 **"a sensation of extreme lassitude":** *Aeronautics*, May 3, 1916, 287.

67 **"tipping his machine":** *Daily Mail* (London), April 14, 1919, 7.

68 **The air forced up:** Walter Raleigh, *The War in the Air: Being the Story of the Part Played in the Great War by the Royal Air Force* (Oxford: Oxford University Press, 1922), 1: 458.

70 **Both were strong-willed:** Muriel Hawker, *H. G. Hawker, Airman: His Life and Work* (London: Hutchinson, 1922), 186.

70 **"we just could not imagine":** *West Australian* (Perth), November 23, 1940, 7.

71 **"extraordinary escape from death":** *The Aeroplane*, July 1, 1914, 17.

71 **"He made light of this":** Miller, *Early Birds*, 11.

CHAPTER 5: CRASHED

72 **"Men and machines were badly needed":** John Alcock, "My Transatlantic Flight," *Badminton Magazine* 52, no. 290 (September 1919): 378.

73 **Only days into the job:** John Alcock to Fred Harvey, February 23, 1915, British Library, Manuscripts Collection: RP 5638/1.

73 **At a performance review:** Royal Naval Air Service, Registers of Officers' Services, The National Archives: ADM 273/7/278.

73 **They described him as an "excellent instructor":** Royal Naval Air Service, Registers of Officers' Services, The National Archives: ADM 273/7/278.

73 **"It was useful work":** Alcock, "My Transatlantic Flight," 378.

73 **In a letter home in early 1915:** Harold Rosher, *In the Royal Naval Air Service: Being the War Letters of the Late Harold Rosher to His Family* (London: Chatto & Windus, 1916), 40.

74 **"Eastchurch was never":** Guy Warner, *World War One Aircraft Carrier Pioneer: The Story and Diaries of Captain J. M. McCleery RNAS/RAF* (Barnsley: Pen & Sword Aviation, 2011), 17.

74 **"One got all that one could":** Alcock, "My Transatlantic Flight," 378.

80 **"I am dead":** *Daily Mirror* (London), March 11, 1919, 9.

81 **She observed that Muller:** Alcock, "My Transatlantic Flight," 379.

81 **"It made my heart leap":** Bishop, *Winged Warfare*, 54–55.

81 **"That pilot never knew":** Bishop, *Winged Warfare*, 163.

81 **The September battle secured:** Alcock, "My Transatlantic Flight," 378.

81 **"most praiseworthy balance":** Report on action of morning of 30 September 1917, prepared October 1, 1917, The National Archives: ADM 137/2197.

82 **Addressing Britain's Parliament:** Hansard: HC Deb, November 19, 1918, vol. 110, cc3318–3323.

82 **"a Christmas which some of us":** Alcock, "My Transatlantic Flight," 382.

83 **"not at all fit":** Arthur Whitten Brown to the secretary of the War Office, December 26, 1918, The National Archives: WO 339/27707.

PART TWO

87 **"Aviators do not know":** *Pall Mall Gazette* (London), July 6, 1912, 12.

CHAPTER 6: SLEEP AWAY THE NEXT TWO MONTHS

89 **"One evening, soon after the Armistice":** Muriel Hawker, *H. G. Hawker, Airman*, 198.

90 **"Why should you think":** Hawker, *H. G. Hawker*, 199.

91 **"It wasn't a question":** Bramson, *Pure Luck*, 94.

91 **"I am getting my kick in":** *The Herald* (Melbourne), May 20, 1919, 9.

93 **The reports don't explain how:** *Western Evening Herald* (Plymouth, UK), September 22, 1913, 2.

94 **"All the pilots will tell you":** *Evening Post* (Wellington, NZ), April 28, 1919, 9.

95 **"as fresh as paint":** *Daily Sketch* (London), April 2, 1919, 24.

95 **"In the past Newfoundland has":** *St. John's Daily Star*, April 12, 1919, 1.

96 **"a brown, bleak and rugged barrier":** Lord Birkenhead, *The Story of Newfoundland* (London: Horace Marshall & Son, 1920), 11.

97 **"I longed to sleep":** Hawker, *H. G. Hawker, Airman*, 199.

CHAPTER 7: AS CLOSE AS A CLAM

98 **After naming a few notable:** *St. John's Daily Star*, March 28, 1919, 1.

100 **"the last place in which one would look":** Harry Hawker and Kenneth Mackenzie Grieve, *Our Atlantic Attempt* (London: Methuen, 1919), 39.

101 **"Captain Fenn certainly got the very best":** Hawker and Grieve, *Our Atlantic Attempt*, 39.

103 **A newspaper reporter at St. John's:** *Daily Mail* (London), April 8, 1919, 5.

104 **"We just wanted to get the thing":** Hawker and Grieve, *Our Atlantic Attempt*, 41–42.

104 **For eighteen-year-old reporter:** *Evening Telegram* (St. John's), August 25, 1919, 11.

105 **Hawker and Grieve experimented:** Joey Smallwood, *The New Newfoundland* (New York: Macmillan, 1931), 196.

105 **"Hawker and Mackenzie-Grieve are not communicative":** *St. John's Daily Star*, April 11, 1919, 1.

106 **"as close as a clam":** *St. John's Daily Star*, April 12, 1919, 8.

CHAPTER 8: ALL SORTS OF STUNTS

108 **"No sooner had we struck":** *The People* (London), November 1, 1914, 9.

110 **A church magazine later observed:** *New Forest Magazine*, January 1915, quoted in Stanley Lane, *The Story of St. Andrew's Church, Tiptoe* (Christchurch, UK: Natula Publications, 2005), 59.

111 **The blast felt by those on board:** *Newcastle Daily Chronicle*, November 3, 1914, 7.

111 **The gun was reportedly fired:** *The Aeroplane*, June 18, 1924, 531.

113 **"Although young in years":** *St. John's Daily Star*, April 14, 1919, 10.

114 **"I am afraid these life saving":** *New York Times*, April 30, 1919, 9.

114 **"cross the Atlantic, not fall":** *Surrey Advertiser*, April 19, 1919, 4, and *New York Times*, April 15, 1919, 3.

115 **"would discuss the machine":** *Evening Telegram* (St. John's), August 25, 1919, 11.

115 **"one of those English sports":** *St. John's Daily Star*, April 14, 1919, 1.

116 **"He is enthusiastic for the venture":** *The Times* (London), March 25, 1919, 10.

117 **"The conditions for flying":** *Daily Mail* (London), March 27, 1919, 5.

118 **"As if luck were deliberately":** *New York Times*, April 12, 1919, 1.

119 **As officials, mortified, tried:** *New York Times*, April 12, 1919, 4.

120 **"The two lean, thin-faced, clear-eyed":** *New York Times*, April 13, 1919, 1, 3.

121 **Doing so blind would:** *New York Times*, April 13, 1919, 3.

121 **"The machine was in readiness":** *St. John's Daily Star*, April 14, 1919, 10.

121 **"The wind is doing all sorts":** *St. John's Daily Star*, April 14, 1919, 10.

122 **"It was a beautiful rise":** *New York Times*, April 18, 1919, 4, and *Evening Telegram* (St. John's), April 17, 1919, 6.

CHAPTER 9: DO BUCK UP AND START

125 **He studied at Stubbington House:** *Portsmouth Evening News*, April 8, 1960, 2.

125 **On making the appointment:** Nicholas Lambert, *Sir John Fisher's Naval Revolution* (Columbia: University of South Carolina, 1999), 290.

126 **"The official and social world":** Mark Kerr, *Land, Sea, and Air: Reminiscences of Mark Kerr* (London: Longmans, Green, 1927), 279.

126 **"My principal qualification":** Kerr, *Land, Sea, and Air*, 298.

127 **"We are not interested in short":** *Daily Mail* (London), April 9, 1919, 5.

127 **"no superior as a big-machine pilot":** Kerr, *Land, Sea, and Air*, 300.

129 **"At one time we considered":** Kerr, *Land, Sea, and Air*, 299.

129 **One reporter, praising Reid's work:** *New York Times*, April 19, 1919, 1.

130 **"The population is so eager":** *New York Times*, April 19, 1919, 1.

131 **"It was like tearing up":** *Evening Telegram* (St. John's), May 19, 1919, 7.

132 **"It was a big strain":** *New York Times*, May 23, 1919, 3.

132 **Both feared the consequences:** *Daily Mail* (London), April 24, 1919, 5.

132 **Sensing a shift in mood:** *Daily Mail* (London), April 21, 1919, 6.

133 **"For sheer agony waiting on":** Hawker and Grieve, *Our Atlantic Attempt*, 41.

134 **Bell Island ore was so abundant:** P. T. McGrath, *Newfoundland in 1911* (London: Whitehead, Morris, 1911), 107.

134 **This didn't bother Hawker:** Hawker and Grieve, *Our Atlantic Attempt*, 43.

134 **Its advertisements in the local press:** *Evening Telegram* (St. John's), January 24, 1919, 1.

135 **"the most up-to-date":** *St. John's Daily Star*, March 4, 1919, 10.

135 **"Do buck up and start":** *New York Times*, April 23, 1919, 6.

137 **This task appeared:** *Evening Telegram* (St. John's), May 19, 1919, 7.

138 **"difficulties arose which could not":** Kerr, *Land, Sea, and Air*, 301.

CHAPTER 10: HANG THE WEATHER!

139 **"in the game from the very jump":** *Daily Mail* (London), June 16, 1919, 4.

139 **"It was not that he wanted":** *The Aeroplane*, June 18, 1919, 2393.

141 **As the three friends caught up:** *Manchester Guardian*, June 16, 1919, 7.

142 **"I caught neither a Scotch":** Brown, *Flying the Atlantic in Sixteen Hours*, 6.

143 **"While I was talking with the superintendent":** Brown, *Flying the Atlantic in Sixteen Hours*, 10–11.

144 **"There was shellac":** Iain Carson, "Vimy Veterans," *Vickers News*, May 23, 1969, 5.

144 **"I saw it take off":** Carson, "Vimy Veterans," 5.

145 **The captain also gave his younger:** Brown, *Flying the Atlantic in Sixteen Hours*, 12.

145 **For his part, Rostron remembered:** Arthur Rostron, *Home from the Sea* (London: Cassell, 1931), 180.

146 **The answer they got back:** Brown, *Flying the Atlantic in Sixteen Hours*, 17.

147 **"Mr. Harry Hawker's machine":** *Daily Telegraph* (London), April 15, 1919, 9.

147 **Hawker and Raynham felt like:** *The Times* (London), May 19, 1919, 14.

148 **Morgan, noting the string:** *New York Times*, May 7, 1919, 16.

149 **"There is a strong feeling here":** *New York Times*, May 12, 1919, 4.

149 **The correspondent for Lord Northcliffe's:** *Daily Mail* (London), May 13, 1919, 7.

150 **"After that there was only one thing":** Hawker and Grieve, *Our Atlantic Attempt*, 46.

151 **"Hang the weather!":** *Manchester Guardian*, May 19, 1919, 7.

151 **"I have a perfect machine":** *Irish Independent* (Dublin), May 20, 1919, 5.

151 **"Hawker, as usual, was just":** Kerr, *Land, Sea, and Air: Reminiscences of Mark Kerr*, 302.

151 **Grieve, addressing friends waiting:** *Irish Independent* (Dublin), May 20, 1919, 5.

151 **With the Rolls-Royce engine:** *Yorkshire Telegraph and Star*, May 19, 1919, 6.

153 **"Every lip gave expression":** *St. John's Daily Star*, May 19, 1919, 1.

153 "All the people of this country": *New York Times*, May 19, 1919, 16.

154 "While London slept, Mr. Hawker": *Daily Telegraph* (London), May 19, 1919, 11.

154 "Hawker and Grieve are among": *St. John's Daily Star*, May 19, 1919, 1.

154 "The great adventure": *Daily Mail* (London), May 19, 1919, 7.

154 "The general feeling to-night is anxiety": *Manchester Guardian*, May 19, 1919, 6.

154 "By the time this appears": *Evening Telegram* (St. John's), May 19, 1919, 6.

CHAPTER 11: IS RAYNHAM ALRIGHT?

155 "Both were in high spirits": *St. John's Daily Star*, May 19, 1919, 1.

155 "He was just a schoolboy": *St. John's Daily Star*, May 27, 1919, 1.

156 "The school boys all knew him": *St. John's Daily Star*, May 19, 1919, 1.

156 "rather reticent and disinclined": *Evening Telegram* (St. John's), August 25, 1919, 11.

157 "A strong bond of affection": *St. John's Daily Star*, May 23, 1919, 13.

160 "Is Raynham alright?": *St. John's Daily Star*, May 27, 1919, 1.

160 By then, Raynham had turned: *New York Times*, May 19, 1919, 3.

160 "Poor old machine": *St. John's Daily Star*, May 27, 1919, 1.

161 "altogether unsuitable for a 'take-off'": Brown, *Flying the Atlantic in Sixteen Hours*, 18.

162 "This hope will die": *New York Times*, May 20, 1919, 2.

CHAPTER 12: NEVER GIVE UP HOPE

164 "If, as now seems apparent": *New York Times*, May 20, 1919, 16.

164 "I have come here to get away": *New York Times*, May 21, 1919, 1.

165 "Perhaps I shall not hear": *Daily Mail* (London), May 20, 1919, 7.

166 "If things don't go quite right": Hawker, *H. G. Hawker, Airman*, 254.

166 "It is over": *New York Times*, May 20, 1919, 2.

166 Fred Raynham tried to remain optimistic: *Daily Mail* (London), May 21, 1919, 7.

166 A message on Tuesday the 20th: *New York Times*, May 22, 1919, 4.

167 "How else is it": *New York Times*, May 22, 1919, 4.

167 "I have not by any means": *Daily Mail* (London), May 21, 1919, 7.

167 "Hope deferred—but still hope": *Daily Mail* (London), May 21, 1919, 6.

167 "To this day": *New York Times*, May 21, 1919, 1.

168 A report cabled to New York: *New York Times*, May 22, 1919, 4.

168 "was a gamble with their luck": *Flight*, May 22, 1919, 661.

168 "There are at least 50 Scandinavian": *Daily Mail* (London), May 22, 1919, 7.

168 A columnist writing in the same issue: *Daily Mail* (London), May 22, 1919, 6.

169 "It seemed that conceivably": *Daily Telegraph* (London), May 26, 1919, 9.

169 "The place where he is supposed": *Daily Mail* (London), May 24, 1919, 5.

170 "The position of this storm": *Daily Mail* (London), May 24, 1919, 5.

CHAPTER 13: NOT YET FAVORABLE—BUT POSSIBLE

172 **Hawker must have reflected:** *New York Times*, May 19, 1919, 2.

CHAPTER 14: YOUR SUDDEN AND TRAGIC SORROW

178 **They received countless telegrams:** *Daily Mail* (London), May 29, 1919, 6.
179 **"You're looking down in the mouth":** *Daily Mail* (London), May 26, 1919, 7.
179 **But her own local newspaper:** *Surrey Advertiser*, May 24, 1919, 5.
179 **"Whenever the time comes":** *Daily Mail* (London), May 27, 1919, 7.
179 **"The King, fearing the worst":** *Sheffield Daily Telegraph*, May 26, 1919, 5.
180 **"Although there is still just a hope":** *Sunday Times* (London), May 25, 1919, 8.
180 **"Hawker search abandoned":** *Sunday Times* (London), May 25, 1919, 12.
180 **As they drove back:** Brown, *Flying the Atlantic in Sixteen Hours*, 15.
180 **But as the week went on:** Brown, *Flying the Atlantic in Sixteen Hours*, 21.
182 **At the end of a rueful letter:** *New York Times*, May 23, 1919, 3.
183 **He might have been more reserved:** *Daily Mail* (London), May 28, 1919, 5.
186 **"There is no feverish anxiety":** *Daily Mail* (London), May 27, 1919, 6.
187 **"Our delight knew no bounds":** *The Scotsman* (Edinburgh), May 27, 1919, 5.

CHAPTER 15: DON'T CRY

188 **"We were within about twenty feet":** Hawker and Grieve, *Our Atlantic Attempt*, 99–100.
188 **"I gave her a good mouthful":** Hawker and Grieve, *Our Atlantic Attempt*, 55.
189 **"We flew around for some two hours":** Hawker and Grieve, *Our Atlantic Attempt*, 56.
190 **"It was like being in a small motor-boat":** *Daily Mail* (London), May 28, 1919, 5.
190 **"Suddenly," Hawker remembered, "a hull loomed":** Hawker and Grieve, *Our Atlantic Attempt*, 56–57.
193 **"They were pleasant fellows":** *Birmingham Daily Gazette*, May 30, 1919, 1.
193 **"Neither Grieve nor myself":** Hawker and Grieve, *Our Atlantic Attempt*, 58.
193 **"They are a class apart":** *Daily Mail* (London), May 27, 1919, 6.
194 **"I am very pleased":** *Leicester Daily Post*, June 2, 1919, 1.
194 **They were given what Hawker:** *Daily Mail* (London), May 27, 1919, 7.
194 **"If Scapa Flow presented":** Sydney Fremantle, *My Naval Career 1880–1928* (London: Hutchinson, 1949), 273.
194 **"From the moment of your departure":** *Dundee Courier*, May 27, 1919, 5.
195 **"Smile, Grieve, smile":** *Daily Mail* (London), May 28, 1919, 5.
195 **"riot of human emotion":** *Daily Mail* (London), May 28, 1919, 5.
195 **"The telephone was at work":** *Daily Mail* (London), May 26, 1919, 7.
195 **"All well, making for home":** *Surrey Advertiser*, May 28, 1919, 3.
195 **"Safe. Will wire later":** *Daily Mail* (London), May 26, 1919, 7.
195 **"She is a most wonderful woman":** *Sheffield Evening Telegraph*, May 26, 1919, 3.
196 **"I can't remember exactly":** *Daily Mail* (London), May 26, 1919, 7.

196 **Muriel returned to a telegram:** *The Globe* (London), May 26, 1919, 1.

196 **Lord Northcliffe, who had been so moved:** *Daily Mail* (London), May 27, 1919, 7.

197 **Accompanied to the cinema:** Hawker, *H. G. Hawker, Airman*, 257.

197 **"I have been loyally backed up":** *The Scotsman* (Edinburgh), May 29, 1919, 7.

197 **"He just said the sweetest":** Hawker, *H. G. Hawker, Airman*, 266.

197 **Vast crowds met the aviators:** *Daily Mail* (London), May 28, 1919, 5.

197 **"Mr. Hawker stepped":** *Daily Mail* (London), May 28, 1919, 5.

199 **"As one family we have come":** *Daily Mail* (London), May 30, 1919, 8.

199 **Describing the endeavour:** Reprinted in *Evening Telegram* (St. John's), May 27, 1919, 6.

CHAPTER 16: MADDENING

200 **"It was in men's eyes":** *New York Times*, May 26, 1919, 2.

200 **"It always did one good":** Kerr, *Land, Sea, and Air*, 302.

201 **"For Grieve, as an expert":** Brown, *Flying the Atlantic in Sixteen Hours*, 13.

202 **Others leaned nonchalantly:** Brown, *Flying the Atlantic in Sixteen Hours*, 23.

203 **When he got back to St. John's:** Brown, *Flying the Atlantic in Sixteen Hours*, 24.

204 **"Both aircraft, respectively the largest":** *New York Tribune*, May 28, 1919, 2.

205 **"This means a fight":** *New York Times*, May 29, 1919, 2.

206 **At Harbour Grace that morning:** *Daily Mail*, London, July 4, 1919, 6.

206 **In later life, he recalled:** Bob Dicker, "Some More about Brooklands—By Request," *Official Journal of the Vintage Motor Cycle Club*, February 1979, 49.

207 **"Under Alcock's skillful hands":** Brown, *Flying the Atlantic in Sixteen Hours*, 27.

208 **"Alcock's always ruddy face":** *New York Evening World*, June 10, 1919, 8.

208 **"I am perfectly satisfied":** *Daily Mail* (London), June 10, 1919, 5.

208 **At the post office, he cabled:** *Daily Mail* (London), June 11, 1919, 5.

210 **The giant machine took off:** *Evening Telegram* (St. John's), June 12, 1919, 6.

210–211 **Brown later described Maxwell-Muller:** Brown, *Flying the Atlantic in Sixteen Hours*, 29.

212 **Alcock was exasperated:** *Daily Mail* (London), June 14, 1919, 5.

213 **Brackley noted his conclusion:** Frida H. Brackley, *Brackles: Memoirs of a Pioneer of Civil Aviation* (London: Simpkin Marshall, 1952), 40–41.

213 **"the worst blow that could have happened":** Kerr, *Land, Sea, and Air*, 303.

CHAPTER 17: A HURRIED LINE BEFORE I START

216 **"We would toss a handful":** *New York Times*, June 16, 1919, 3.

217 **"My Dear Elsie":** "Philatelic Rarities: Newfoundland: Letter from John Alcock to his Sister dated 12th June, 1919 and Carried in the Cover Above," British Library Online Gallery, www.bl.uk/onlinegallery/onlineex/philrar/n/010ca0nl0001919u00001002.html.

217 **"It always gave me a lot":** Edwin C. Parsons, *I Flew with the Lafayette Escadrille* (Indianapolis: E. C. Seale, 1963), 229.

218 **His last words to her:** *Daily News* (London), June 16, 1919, 1.

219 **"It's a long flight":** *Daily Mail* (London), June 16, 1919, 5.

219 **Then the pair said their closing words:** *New York Times*, June 15, 1919, 2.

220 **"Oh, well, if we crash":** *New York Times*, June 16, 1919, 3.

220 **"Bob, what are you doing":** Dicker, "Some More about Brooklands—By Request," 50.

221 **A cry went up:** *New York Times*, June 15, 1919, 1.

222 **"I'm going to find them":** Dicker, "Some More about Brooklands—By Request," 50.

222 **"Look Bob":** Dicker, "Some More about Brooklands—By Request," 50.

222 **"It looked bloody marvellous":** Carson, "Vimy Veterans."

223 **Before leaving the land behind:** Brown, *Flying the Atlantic in Sixteen Hours*, 41.

PART THREE

225 **"We don't yet know how far":** Hewlett, *Old Bird*, 174.

CHAPTER 18: WE MUST GET STARS

227 **The aeroplane had entered:** Alcock, "My Transatlantic Flight," 384.

233 **"We must get stars":** Brown, *Flying the Atlantic in Sixteen Hours*, 50 and 53.

234 **"An aura of unreality":** Brown, *Flying the Atlantic in Sixteen Hours*, 57.

CHAPTER 19: A BAD LANDING

237 **"By the time we reached Vitry":** Holland, "Some of My Many Thrills," 48.

238 **"Automatically," the pilot wrote:** "Wing Adjutant" [Wilfred Blake], *Royal Flying Corps*, 24.

238 **Shells from the ground came:** "Wing Adjutant" [Wilfred Blake], *Royal Flying Corps*, 55.

239 **Brown was spared this horror:** Holland, "Some of My Many Thrills," 48.

239 **"We made a bad landing":** Holland, "Some of My Many Thrills," 48.

239 **"I had my leg broken":** Arthur Whitten Brown to the secretary of the War Office, October 11, 1917, The National Archives: WO 339/27707.

240 **"I next woke up":** Holland, "Some of My Many Thrills," 48.

240 **"I knew the deadly heart-sickness":** Brown, *Flying the Atlantic in Sixteen Hours*, 9.

240 **In May 1916, a naval admiral:** Hansard: HL Deb, May 31, 1916, vol. 22, cc249–68.

241 **"They don't hurt you any more":** Arthur Whitten Brown to his parents, March 25, 1916, Imperial War Museum Archive: Documents.3926, box no. 84/22/1.

241 **"When we reached Clausthal":** *Dumfries and Galloway Standard*, July 13, 1918, 5.

241 **In one parliamentary debate:** Hansard: HL Deb, May 31, 1916, vol. 22, cc249–68.

242 **One pilot imprisoned at Clausthal:** Sir William Hugh Stobart Chance, C.B.E., "Part 6—Prisoner of War at Clausthal (1917–18)," September

1970, *The Worcestershire Regiment*, www.worcestershireregiment.com/wr .php?main=inc/whs_chance_7.

244 **"I wrote you from Constance":** J. Harvey Douglas, *Captured: Sixteen Months as a Prisoner of War* (New York: George H. Doran, 1918), 155–56.

244 **"We were examined":** Douglas, *Captured*, 153.

245 **For four weeks, the prisoners:** "POW diaries—Captain Percival Lowe," *Europeana 1914–1918*, no date, 81, http://embed.europeana1914-1918.eu/en/ contributions/3963.

245 **"We were informed that this delay":** "Report by Second Lieutenant H. G. Frost, 9th Suffolk Regiment (Attached Royal Flying Corps)," May 27, 1917, The National Archives: WO 161/96/16.

245 **"There wasn't a cheer":** Douglas, *Captured*, 157.

245 **"In a very short time I saw":** Lowe, "Diary," 86.

246 **"We looked out of the windows":** Douglas, *Captured*, 158.

246 **"He will march, surrounded":** "F. B. B." (former British officer prisoner of war), statement, ca. December 1917, The National Archives: FO 383/398.

246 **"the most ghastly butcher":** "Report on the Treatment by the Enemy of British Officers, Prisoners of War, in Camps Under the 10th (Hanover) Army Corps," ca. June 1918, The National Archives: FO 383/399.

246 **Niemeyer once gathered his guards:** "Report on the Treatment by the Enemy of British Officers . . ."

247 **Then he turned to another of the men:** "Report on the Treatment by the Enemy of British Officers . . ."

247 **About seventy years old:** "Report on the Treatment by the Enemy of British Officers . . ."; H. G. Durnford, *The Tunnellers of Holzminden (With a Side-Issue)* (Cambridge: Cambridge University Press, 1920), 18.

247 **He shouted at the young soldier:** Major A. E. Haig (former British officer prisoner of war), statement, January 28, 1918, The National Archives: FO 383/398.

247 **A report prepared for a British government:** "Report on the Treatment by the Enemy of British Officers . . ."

247 **A member of the Royal Flying Corps:** Percy Edwin Butcher, *Skill and Devotion: A Personal History of the Famous No. 2 Squadron of the Royal Flying Corps* (Middlesex: Radio Modeller Book Division, 1971), 46.

248 **"If one had not wanted":** Holland, "Some of My Many Thrills," 48.

248 **"It was as a prisoner of war":** Brown, *Flying the Atlantic in Sixteen Hours*, 10.

248 **A secret British government memorandum:** Memorandum, August 1918, The National Archives: WO 383/418.

CHAPTER 20: THE ROTTENEST LUCK IN THE WORLD

251 **"It appeared as if we could stretch":** Brown, *Flying the Atlantic in Sixteen Hours*, 63.

252 **"The complete absence of such contact":** Brown, *Flying the Atlantic in Sixteen Hours*, 68.

254 **Stanley Wise later recalled flying:** Stanley Wise, "My Most Thrilling Flight: Turkish Delight," *Popular Flying*, August 1934, 241.

256 **"With a red mist in front":** Wise, "My Most Thrilling Flight," 242.

257 **Then Alcock and his crewmates:** Alcock, "My Transatlantic Flight," 380.

258 **"The food was scarce and bad":** Alcock, "My Transatlantic Flight," 380.

258 **"There were cries and shouts":** Francis Yeats-Brown, *Caught by the Turks* (London: Edward Arnold, 1919), 180–81.

259 **"One was utterly lost":** Yeats-Brown, *Caught by the Turks*, 182.

259 **"a punishment as barbarous":** Yeats-Brown, *Caught by the Turks*, 187.

259 **"Captivity is a minor form":** Yeats-Brown, *Caught by the Turks*, 98.

259 **"I had the rottenest luck":** Jack Alcock to Alfred Alcock, October 15, 1917, Science and Industry Museum Archive: YA1991.437/MS0066/1.

260 **Years later, he wrote how Alcock's:** Wise, "My Most Thrilling Flight," 268.

260 **"Physical courage enables us":** Personal correspondence, Tony Alcock to David Rooney, June 13, 2023.

260 **"Well old boy":** Jack Alcock to Fred Moseley, postcard, January 23, 1918, Science and Industry Museum Archive: YMS0063.

261 **"There we had a good time":** Alcock, "My Transatlantic Flight," 380.

261 **"Tell all the boys to write":** Jack Alcock to James Sinclair, July 14, 1918, Imperial War Museum Archive: Documents.1033.

261 **A fellow Kedos prisoner recalled:** Tony Spackman, ed., *Captured at Kut: Prisoner of the Turks—The Great War Diaries of Colonel W. C. Spackman* (Barnsley: Pen & Sword Military, 2008), 171.

262 **"Kedos had ceased to exist":** Charles Leonard Woolley, ed., *From Kastamuni to Kedos: Being a Record of Experiences of Prisoners of War in Turkey, 1916–1918* (Oxford: Basil Blackwell, 1921), 102.

262 **"a town of the dead":** *Western Times* (Exeter, UK), December 2, 1918, 4.

262 **"None of us will ever forget":** Alcock, "My Transatlantic Flight," 381.

262 **"We lived in the open":** Alcock, "My Transatlantic Flight," 381.

263 **"He told me in Kedos":** Spackman, *Captured at Kut*, 172.

CHAPTER 21: WE DID NOT LET YOU DOWN

265 **"Alcock grabbed my shoulder":** Brown, *Flying the Atlantic in Sixteen Hours*, 74.

265 **Brown scribbled a short note:** Alcock and Brown, *Our Transatlantic Flight*, 81.

267 **Once they were safely away:** *Daily Mail* (London), June 16, 1919, 5.

267 **An officer shouted out:** Brown, *Flying the Atlantic in Sixteen Hours*, 79–80.

267 **They also cabled the mechanics:** John Alcock and Arthur Whitten Brown to Vickers mechanics, telegram, received June 16, 1919, Science and Industry Museum Archive: YMS0063.

268 **"Naturally enough, they amplified":** James Lansdale Hodson, "Interviewing the High and Low," *The Listener* 11, no. 260 (January 3, 1934): 16.

268 **"We came down quickly":** James Lansdale Hodson, "First Aeroplane across the Atlantic," *The Listener* 19, no. 493 (June 23, 1938): 1344.

269 "I answered questions mechanically": Brown, *Flying the Atlantic in Sixteen Hours*, 82.

269 "We are too near it": *New York Times*, June 16, 1919, 1.

269 His mind showed him visions: Brown, *Flying the Atlantic in Sixteen Hours*, 83.

270 "Thanks very much, old chap": *Daily Mail* (London), June 17, 1919, 6.

270 "I didn't like it much": *Daily Mail* (London), June 18, 1919, 5.

270 "Teddy's safe. He's done it": *Dundee Evening Telegraph*, June 16, 1919, 5.

270 "I regarded it as a lucky omen": *Daily Mirror* (London), June 16, 1919, 2.

271 People shouted after them: *Daily Mail* (London), June 18, 1919, 5.

272 "Mate, we've done it": Carson, "Vimy Veterans."

272 "Bob," he told Dicker: Dicker, "Some More about Brooklands—By Request," 51.

273 For a while, all Mary Alcock: *Manchester Evening News*, June 16, 1919, 4.

273 The idea of a knighthood: *The Aeroplane*, June 25, 1919, 2552.

274 "a most happy augury": *Daily Mail* (London), June 16, 1919, 5.

274 "In each case," he said: Brown, *Flying the Atlantic in Sixteen Hours*, 4.

275 "The thought of Alcock": *New York Times*, June 22, 1919, 16.

275 One of Alcock's fellow aviators: *The Aeroplane*, June 18, 1919, 2395.

276 Reporters said he got to the door: *New York Times*, July 30, 1919, 9.

276 "I wouldn't do it again": Rostron, *Home from the Sea*, 181.

276 After landing in Ireland: *Poverty Bay Herald*, July 16, 1919, 3.

277 He described it as a "binge": Jack Alcock to James Sinclair, October 27, 1919, Imperial War Museum Archive: Documents.1033.

277 The best that local reporters: *St. John's Daily Star*, June 16, 1919, 1.

278 He said such a move: *New York Times*, June 20, 1919, 15.

278 "no useful purpose would be served": Herbert Brackley, "Newfoundland to New York, 1919," *The Aeroplane*, May 4, 1938, 533.

278 "It was a bitter disappointment": Kerr, *Land, Sea, and Air*, 303.

280 Raynham waved his hand: *New York Times*, July 18, 1919, 5.

EPILOGUE

285 When the display was formally opened: "Notes for the Minister's Speech on Opening the Aeronautic Gallery at the Science Museum on 10th July," 1963, Science Museum Archive, London: Z183/2.

286 When Charles Lindbergh landed: *Dundee Evening Telegraph*, June 14, 1929, 2.

286 Amelia Earhart, who flew the Atlantic: *Western Star* (Curling, NL), November 22, 1933, 1.

286 Amy Johnson, after her solo journey: *Hartlepool Northern Daily Mail*, June 3, 1930, 5.

287 "tall lighthouses on the path of progress": *Daily Mail* (London), June 17, 1919, 5.

CREDITS

val collection of trans-Atlantic flight materials, 1919: Sopwith hangar, Mt. Pearl, St. John's, showing case containing trans-Atlantic aeroplane.

109 *Illustrated War News*, November 4, 1914.

117 The Rooms Provincial Archives Division, E 70-49, T. B. Hayward collection.

123 Archives & Special Collections, Queen Elizabeth II Library, Memorial University Libraries, St. John's, NL, COLL-505 Margaret A. Carter archival collection of trans-Atlantic flight materials, 1919: Views taken by Raynham on trial flight at St. John's, Martinsyde trans-Atlantic aeroplane, May 1919, Conception Bay.

131 The Rooms Provincial Archives Division, VA 67-1.2, Kerr-Brackley photograph collection.

133 The Rooms Provincial Archives Division, VA 140-50, Hawker and Mackenzie-Grieve album.

136 The Rooms Provincial Archives Division, E 67-46, T. B. Hayward collection.

138 The Rooms Provincial Archives Division, VA 67-6.1, Kerr-Brackley photograph collection.

149 Library of Congress, Prints & Photographs Division, LC-DIG-ggbain-28813.

152 The Rooms Provincial Archives Division, VA 140-47, Hawker and Mackenzie-Grieve album.

156 The Rooms Provincial Archives Division, A 16-95, Philip and Kathleen Hood collection.

158 The Rooms Provincial Archives Division, E 70-19, T. B. Hayward collection.

182 The Rooms Provincial Archives Division, E 70-20, T. B. Hayward collection.

196 Muriel Hawker, *H. G. Hawker, Airman: His Life and Work* (Hutchinson, 1922).

198 Library of Congress, Prints & Photographs Division, LC-DIG-ggbain-28991.

211 The Rooms Provincial Archives Division, E 69-60, T. B. Hayward collection.

218 Chronicle / Alamy Stock Photo.

223 © Science & Society Picture Library / Getty Images.

249 H. G. Durnford, *The Tunnellers of Holzminden (With a Side-Issue)* (Cambridge University Press, 1920).

266 *Philips' Handy Atlas of the Counties of Ireland* (George Philip & Son, 1881).

268 Chronicle / Alamy Stock Photo.

271 *The Silver Jubilee Book: The Story of 25 Eventful Years in Pictures* (Odhams Press, 1935).

274 Chronicle / Alamy Stock Photo.

276 Library of Congress, Prints & Photographs Division, LC-DIG-ggbain-29466.

281 The Rooms Provincial Archives Division, A 16-88, Philip and Kathleen Hood collection.

282 City of St. John's Archives, Photo # 11-05-1488.

284 David Rooney.

286 David Rooney.

288 Library of Congress, Prints & Photographs Division, LC-DIG-hec-47465.

INDEX